Advance Praise for The Accessible Home

"*The Accessible Home* stands as the roadmap to the universal home."
—Sam Maddox, Knowledge Manager, The Christopher and Dana Reeve Foundation

"Inspiring, visionary, and yet totally down to earth. This is a must-read for anyone designing the home of a lifetime."
—Hugh Delehanty, co-author of *Caring for Your Parents: The Complete Family Guide*

"*The Accessible Home* is a comprehensive approach to designing living spaces for people with disabilities from an individualized perspective."
—Tim Gilmer, Editor, *New Mobility* magazine

"An inspired, beautiful, and practical book...this is the most complete home design book I've ever seen."
—Elaine Ostroff, Hon. AIA, Founding Director, Institute for Human Centered Design

"This book shows us that an accessible house can incorporate beautiful design and meet the functional needs of the people who live there through their lifespan."
—Brenda Battat, Executive Director, Hearing Loss Association of America

"In *The Accessible Home*, Deborah Pierce delivers abundant proof that the dream is very much available for anyone with a functional limitation."
—Valerie Fletcher, Executive Director, Institute for Human Centered Design

"Architect Deborah Pierce dispels many misconceptions and explains clearly how accessible homes are elegant, more functional, and safer for people of all ages and abilities."
—Laura Montllor, Architect, AIA, Executive Director, HomeFreeHome.org

"This book is a gem. It provides an excellent framework for the design of living spaces that accommodate the practical needs of people with various special needs in a dignified, functional, and stylish manner."
—Eileen Komanecky, Founder, simpleHome, technology resources for smart homes

the
accessible
home

DESIGNING FOR ALL
AGES AND ABILITIES

Deborah Pierce

The Taunton Press

Text © 2012 by Deborah Pierce, AIA
Photographs © 2012 by Kathy Tarantola, except where noted
Illustrations © 2012 by The Taunton Press, Inc.
Front cover photograph: Kathy Tarantola;
     architect: LDa Architecture & Interiors

The Taunton Press
Inspiration for hands-on living®

The Taunton Press, Inc.,
63 South Main Street,
PO Box 5506, Newtown, CT 06470-5506
e-mail: tp@taunton.com

Editor: Peter Chapman
Copy editor: Seth Reichgott
Indexer: Cathy Goddard
Jacket/Cover design: Alison Wilkes
Interior design: Teresa Fernandes
Layout: Laura Lind Design
Illustrator: Martha Garstang Hill
Photographer: Kathy Tarantola, except where noted

The following names/manufacturers appearing in *The Accessible
Home* are trademarks: Corian®, Washlet®

Library of Congress Cataloging-in-Publication Data in progress

ISBN 978160085-491-0

Printed in the United States of America
10 9 8 7 6 5 4 3 2 1

dedication

DEDICATED TO THE MANY PEOPLE WHO OPENED THEIR HOMES TO ME and shared their stories of making the necessary adjustments to live well with physical limitations. This book would not have been possible without your commitment to create value out of your experience and to allow it to inspire others.

acknowledgments

To the staff at The Taunton Press, thank you for your cooperation at every step of producing this book. To executive editor Peter Chapman, thanks for the many conversations that honed the focus of this book, and also for your deft editorial skills over the course of writing it. Your belief in this architect's foray into authorship has been a contribution beyond words. To photo editors Katy Binder and Erin Giunta and art director Alison Wilkes, thanks for the logistical support in collecting and organizing hundreds of images from dozens of photographers; you've turned an enormous task into a smooth collaboration. To the artists and layout staff at Taunton, thank you for giving this project your best.

To Kathy Tarantola, thank you for your partnership in creating the visual images that accompany my written imagery. Your good-natured patience in adding "just one more view" after a long day of photography has resulted in some truly wonderful pictures. Your warmth and openness to the people in our photos has captured their humanity and helped make the shoots more enjoyable for all. Thanks for bringing your commitment to excellence to each photograph.

To the many photographers whose work fills the book, thank you for taking compelling pictures that allowed me to see the stories each house and homeowner had to tell. To those who donated their photos, your generosity has allowed us to expand the number and quality of images included.

To the architects, designers, and builders whose work is described here in words and pictures, thank you for bringing your creativity to the task of making homes that are both accessible and wonderful to be in. It's a special skill to design a house that is responsive to the unique needs of each homeowner. In overall layouts and in the details, your designs exhibit insight into your clients' needs, and the ways in which the built environment can really make a contribution to people's lives.

To the engineers, product representatives, and design specialists who contributed technical knowledge for the sidebars, thank you for sharing your expertise. You remind us that design is a collaborative effort and that success really comes from getting the details right.

To the many people who helped launch this book in various ways, thank you. This includes those who answered Internet surveys, who attended seminars, and who were on hand to literally open doors and lend us a broom on the photo shoots. All these gestures of generosity and kindness have been much appreciated.

A special thank you goes to Etty, who read and commented on the manuscript, and whose insights have strengthened the book. Another goes to Curt, within whose partnership I've honed the skills needed to create responsive and accessible places.

To my friends and family: Thank you for your support and patience while I've been pre-occupied with the book. To my grown children Robin, Casey, and Erika, whose encouragement has given me the confidence I could do this: You've shown by example how exciting life is when we step outside our comfort zone and follow our passions. To Diego: May your world be friendly to individual differences. To Steve: Your belief in me has always been an anchor.

May the ideas in this book spark your own creativity to create new environments and overcome barriers. I look forward to seeing what you do with the ideas in this book.

Deborah Pierce
July 2012

contents

foreword

I have always been interested in the themes of accessibility and domesticity in architecture. Since the early 1960s, when I spent two years at the American Academy in Rome, I have been fascinated by the way that the language of architecture, especially its traditional language, allows people to comprehend the world around them and orient themselves within a city, a building, or a room. Elements such as doors, windows, columns, walls, ceilings, and the ground plane, in various combinations, help differentiate here from there, in and out, public and private. Houses achieve a sense of domesticity through the relationship with their surroundings, the movement of light and the progression of activities throughout the day, and the arrangement of furniture and furnishings that support social activities and create conviviality. To me, architecture and design have always been most accessible when they are legible, when their pragmatic functions and usability are understood intuitively, and when their forms and relationships are not ambiguous.

The front entrance of this Wounded Warrior home combines a generous landing with a spacious porch to invite guests and encourage outdoor gatherings.

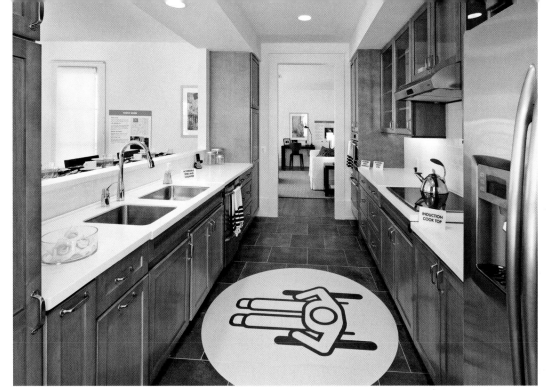

Accessible features in the kitchen of this Wounded Warrior home include an adjustable-height cooktop and an adjustable-height sink with a removable cabinet below. Wide doorways with level floors combine with wheelchair turning space to create an accessible path of travel and work area.

That was before my paralysis. All of the attributes of accessibility and domesticity that I've held dear for so long are still true. However, when I became confined to a hospital bed and then to a wheelchair, I discovered that much of our built environment offers obstacles that range from frustrating to formidable. Many of these would be easy to fix if we architects and designers just paid attention. Some features that permit accessibility are governed by codes and governmental guidelines, and others are simply a matter of common sense.

My firm recently designed several Wounded Warrior homes for the U.S. Army, under a program that provides accessible single-family houses on military bases for disabled servicemen and women and their families. The people who live in the houses for two-year stretches have a wide variety of disabilities—visual, physical, and cognitive—and the designs therefore accommodate many circumstances. As with any house, the relationship between indoors and outdoors, and the organization and character of rooms and passages between them, contributes to its domestic ambiance and to the well-being of each

occupant. However, the small details make it work for a disabled person, or not: the swing of a door, the height of a counter, the placement of a faucet or drain, the shape of the threshold, the contrast of colors, and so on.

So when Deborah Pierce asked me to look at her manuscript for *The Accessible Home* and write the book's foreword, I immediately recognized a kindred spirit, someone who is willing to tackle the small problems along with the large in her quest to make wonderful places where people with disabilities can live comfortably and safely. Her thoughtful case studies are full of helpful ideas, many of them little things that make a world of difference to a disabled person. More than that, however, she sets her sights on creating houses that are beautiful, comfortable, and convivial for the entire family and their guests. That is one of the most gratifying accomplishments for any architect. I commend her on this insightful book, which will prove useful—and accessible—to the public and professionals alike.

Michael Graves, FAIA
Princeton, New Jersey
March 2012

introduction

"Nothing is certain but all things are possible."

—PERUVIAN SAYING

I N THE EARLY 1990S THE AMERICANS WITH DISABILITIES Act was in its infancy, and my architectural practice was busy assessing public buildings for compliance with the ADA. The School Department in my city had embraced a program called "Understanding Handicaps," to teach fourth-graders about human differences through its character-education curriculum. Taught by trained parent facilitators, the program simulated various disabilities so that the students could experience the challenges of being unable to use their bodies fully. Thinking my experience might be of help, I signed up to lead the two-week unit on physical disabilities.

For the last class of the unit we invited a young man to speak with the students about his experiences as a quadriplegic. Jim had broken his spine diving into shallow water. Now living and driving independently, he planned to meet us in the auditorium. The fourth-grade students were waiting for Jim as the minutes ticked by. Recalling a new citywide policy on locking school doors, I went to the front entrance thinking the bell might be out of order. A van was parked beside the curb-cut near the side door, and Jim was seated in his wheelchair at the only school entrance with a ramp—and no doorbell.

Accompanying Jim through the school was an eye-opening experience. The building presented barriers at every turn, starting with the doorbell-less ramped entrance. Corridor doors came in sets of two, narrow heavy oak doors swinging toward the exit that were impossible to enter without assistance. The auditorium floor pitched steeply down toward the stage with narrow maneuvering aisles and landings. The school's only accessible restroom was unisex and child-sized.

Something clicked for me: It would take more than laws to ensure equitable environments, although the ADA was an important start.

Years later in my own practice, a couple turned to me to renovate their house when their daughter Jamie was six. The house was an obstacle course for a child with cerebral palsy and using a power wheelchair, with steep stairs and bathrooms too small for Jamie and an assistant. Her mother suggested I visit Jamie at school to see her in a barrier-free setting. Jamie was a different child there, independent and engaged. It was crystal clear that the environment has a profound impact on who we are and whether we can fulfill our potential as human beings.

WHY I WROTE THIS BOOK

I wrote this book to fill a need that existed when I was designing Jamie's home (see Case Study 16, Suburban Renovation, on p. 174). There was an excellent book available about veterans' homes, but Jamie was a child, and her family's needs were equally important. Medical rehab specialists could recommend adaptive products but not design modifications. Building and access codes that focus on dimensions and clearances gave little guidance in understanding the myriad issues of multiple disabilities. Internet resources are overwhelming and limited, raising the specter of a gadget-filled environment that may be accessible but that also seems cluttered and institutional.

Homeowners are easy prey for a building industry eager to plunk elevator shafts against the walls and cut holes in historic porches so that metal ramps can be installed near the front door. Property values plummet with poorly conceived alterations that solve some problems,

HOW TO USE THIS BOOK

The first part of the book is a review of the accessible home. It is organized by activities rather than rooms, as the open floor plan that characterizes many accessible homes makes traditional rooms obsolete. The second part contains descriptions of 25 accessible homes. These are real homes for real people, living with a wide range of disabilities: conditions from birth, progressive illness, debilitating injuries, hearing and vision loss. There are homes for families, for couples, and for single people. There are new houses and renovations, one loft and two prefabricated modular prototypes, throughout the United States and including Canada and Mexico.

If you are planning a renovation project such as a bathroom, you can start at chapter 6 and then visit the case studies that are referenced there, to see how the bathroom work is integrated into the full home. If you are looking for inspiration, start with the case studies. For ideas, start with the activity centers. Whatever your approach, if you read through the entire book you will begin to see homes in a new light—as places where disabilities disappear and the living is easier for everyone. It's an idea whose time has come.

even as they create others. In talking with architects and homeowners around the country, it's clear that the need still exists for a collective vision of the good life in living with disability. I wrote this book to take accessibility out of its closet.

In writing this book I've spoken with all of the homeowners involved and visited most of the houses featured. What stands out is how beautiful the homes are—not as accessible homes, but as homes that reflect the intelligence and creativity of a collaborative effort between homeowners and their design teams. The homes are sunny and comfortable, both generous and intimate, each a great place to live. This is the possibility of *The Accessible Home*—that when our homes are truly planned with the whole person in mind, we can transcend the ordinary. And in doing so, we are creating a new concept of the American home for the 21st century.

CHAPTER 1

what is an accessible home?

HOME IS WHERE WE PREPARE to meet the world, and where we retreat when the world roughs us up. Homes house our families, our memories, our stuff. The home is our castle—the place we are free to be ourselves. But when disability strikes, that same home can become a prison, presenting barriers, frustrations, and perils at every turn. Living with disability requires constant adaptation, for the person who must learn new skills and for family members who provide support and assistance. Many people hesitate to invest in access upgrades, assuming that

The accessible home has a welcoming entrance: covered and dry, level, sturdy, and ramped.

> *"Disability results from the interaction between persons with impairments and attitudinal and environmental barriers that hinder their full and effective participation in society on an equal basis with others."*
>
> —INTERNATIONAL CLASSIFICATION OF FUNCTIONING, DISABILITY, AND HEALTH

property values will plummet with wheelchair-driven alterations. Or perhaps it's our youth-oriented culture that wants to believe we are all young and fit, and will always be so.

What we all need is a new vision of the home that works for everyone. Thanks to the civil rights activists of the last few decades, this attitude is becoming the new reality. Architects are starting to do exciting things in creating wonderful accessible homes, in partnership with a new generation of people for whom disability is a fact—but not the defining fact—of life. The good news is that a home designed with an understanding of the person's requirements can go a long ways toward making everything easier, for everyone. This book is about how this works.

WHAT IS A DISABILITY?

Disability is a complex phenomenon, an interaction between a person's physical impairments, the activities they need to perform, and barriers presented by the setting in which this occurs. Physical conditions include health problems that can strike at birth, become manifest in youth or middle age, or appear in old age. Activities are any of the tasks involved in everyday life. Barriers include obstacles in the built and natural environment, as well as the social setting (lack of social supports, negative attitudes, inaccessible services and facilities). As either a temporary or permanent condition,

disability affects three out of four people at some time in their lives.

Defining disability as a social and environmental issue, rather than a personal problem for certain individuals, represents a cultural milestone. In treating disability as a continuum rather than categorizing people with disabilities

The accessible home has a generous interior path of travel and rooms with pocket doors.

"Disability is part of the human condition."

—ED ROBERTS

as a separate group, the World Health Organization says that "disability becomes a matter of more or less, not yes or no." These distinctions are critical because they affect all of us. At the same time, they empower us all to ensure that we are creating an environment where people can function effectively.

In the accessible home, living areas are connected visually and spatially.

> FOR MORE ON THIS HOUSE, SEE Case Study 10 (p. 146)

WHY DO WE HAVE ACCESSIBLE HOMES?

Civil Rights legislation starting with the Architectural Barriers Act of 1968 has attempted to equalize opportunities and prevent discrimination on the basis of ability. Energized by returning veterans of the Vietnam War, disability rights activists succeeded in getting passage of the Americans with Disabilities Act of 1990, which set standards for employment practices as well as "places of public accommodation." States and large organizations responded with their own codes, so that today we have an array of standards and regulations that impact how buildings are designed (see the sidebar on the facing page). Single-family homes are usually exempt, although regulations change often and vary by state. It is wise to check local requirements before planning renovations, especially with multi-family apartment buildings, because the trend in the last half-century has been to remove barriers to full participation in all aspects of community life.

WHO NEEDS AN ACCESSIBLE HOME?

The truth is that we all do at some time in our lives, whether for ourselves, family members, or guests. Young families need accessible homes. In the United States, 3 out of 100 babies born each year have significant birth defects, caused by genetics or problems during pregnancy, that result in physical or mental disabilities. Raising a child with a disability impacts family life and requires constant adaptation as the child grows. When designers can provide features that make life easier for young parents and, at the same time, enhance safety and independence for siblings, they serve the changing needs of a family living with disability.

People who become seriously injured need accessible homes. Healthy active adults have a 1 in 4 chance of becoming disabled for at least 3 months at some point in their lives, with the average long-term disability lasting $2\frac{1}{2}$ years. After a fall, accident, or injury we need to relearn everything. People who lose functionality in parts of the body must teach themselves new ways to perform everyday tasks. Each time several steps can be consolidated, life becomes easier. When homeowners can communicate their unique adaptations, and when designers can look at the home in new ways, then the environment can be made compatible with the necessary activities.

People living with illness need accessible homes. Many diseases affect both physical and sensory abilities. For example, Multiple Sclerosis (MS) affects vision as well as muscles, and Amyotrophic Lateral Sclerosis (ALS) affects strength and coordination as well as speech. Progressive illnesses stress partners as they adapt patterns of daily life. By designing places where people can express their abilities as well as cope more easily with disabilities, we create homes that refresh the spirit. By specifying nontoxic building materials, we help boost people's immune systems, reducing the chances of complicating illnesses.

People with sensory limitations need accessible homes. Low vision and hearing loss differ profoundly, but a well-lit and glare-free environment with good sightlines makes it easier for everyone to see and hear someone in the next room and to pay attention in a group. Coping strategies vary depending on when the sensory loss occurs, how quickly it progresses, and the degree of impairment. There is much that can be done to improve acoustics for those with hearing loss and to help with way-finding for those with low

> FOR MORE ON THIS HOUSE, SEE Case Study 7 (p. 131)

The accessible kitchen is comfortably sized. Pull-out shelves expand workspace at activity centers, while low windows put the outdoors in view.

ALPHABET SOUP

UFAS, Section 504, ANSI, ABA, ICC and IBC, the ADA: These acronyms describe regulations enacted to protect the rights of people with disabilities since 1961. And while the guidelines can be found online in easy-to-use formats with clear illustrations, it may be necessary to hire a consultant to determine precisely what jurisdictions impact a project. Workplace and housing requirements vary by industry, ownership, and financing, and so what was done for your neighbor or at your office may not apply to your home. Single-family homes are generally exempt from compliance with access laws, unless they are built by public agencies and sold to individuals, or unless they are in private developments that are open to the public (such as the leasing/sales office).

Dimensions used in this book reference the ADA (Americans with Disabilities Act of 1991), which extended civil rights laws to employment and public places. The 2010 Amendments expanded coverage for multifamily housing (four or more dwelling units). When in doubt about your own compliance requirements, consult with an access codes specialist, as regulations change regularly and vary by state.

Accessible bedrooms can be easily exited in an emergency. Accessible bathrooms are nearby.

vision. A home designed to enhance the sensory experience of its users is easy on the senses for everyone.

People with intellectual limitations need accessible homes. Brain injuries sustained in a stroke or accident impact various skills, depending on where the injury occurs. For example, injuries to the cerebellum affect motor abilities and coordination, whereas injuries to the temporal lobe affect hearing and memory. The number of children diagnosed with learning disabilities is increasing as testing becomes more widespread. Cognitive disabilities caused by aging (dementia) or illness (Alzheimer's) are more common in later years, and medical advances have extended the population's lifespan. By incorporating features that control distraction and enhance safety, designers help everyone function more effectively at home.

People planning to stay put through their later years need accessible homes. AARP surveys report that over 85% of people want to age at home. By age 75, however, it is estimated that 73% of the population will have some functional

limitations. Due to medical advances and healthy lifestyles, the portion of the population over age 55 is predicted to reach 35% by 2050. Most houses are designed around the perceived needs of able-bodied families with young children. Stairways, compact bathrooms and kitchens, 30-in. doors, all-white walls and electrical controls, shiny wood floors and plush carpets all pose hazards for people aging in place. The accessible home challenges popular notions of what a home looks like.

WHY DESIGN AN ACCESSIBLE HOME?

Building or remodeling a home is an investment in time, money, and energy. After being diagnosed with a degenerative illness, or simply in the act of aging, people may find themselves without the resources to forge ahead with renovations. After an injury, the focus is on learning new skills. Home modifications may be off the radar screen. Renovations take time and require the homeowner's attention and good judgment, and so designers do their younger clients a service by showing how well-planned homes can provide payback throughout the life cycle.

During a renovation, many accessibility changes do not add cost; they only require thinking ahead. A carefully planned remodeling job provides ample time to explore alternatives and shop around for the most suitable product. It provides an opportunity to pursue special approvals and obtain competitive bids. Stress levels and change orders are reduced when decisions are not made on the fly. Property values are enhanced with a well-designed home, and the pool of potential buyers increases when that home is also accessible. For renovations at any age, it just makes sense to include accessibility in project goals.

HOW IS AN ACCESSIBLE HOME DIFFERENT FROM OTHER HOMES?

Accessible homes look much like other homes. They often have a sunny open feeling because there are fewer walls between common areas. Doorways are wider and windows taller. Level floors create a comfortable flow between living areas and make rooms easier to keep clean. The kitchen is efficient and ergonomic, with compact storage and broad, sleek work areas. Bathrooms are a little more spacious than in traditional homes. Air quality is good, with operable windows and attention to nontoxic building materials. Sound quality is also good, thanks to simple acoustic features that block unwanted noise and enhance communication. Electrical fixtures provide even and glare-free illumination. Finish materials and details are selected for safety as well as general utility and attractiveness. The house is safe, designed to reduce falls. It is low-maintenance —well designed, durable, and skillfully constructed. The accessible home is the home of the future, but it is also the way we want to live now.

The accessible home has multipurpose rooms that expand and contract to accommodate different types of gatherings. Serving is easy because dining areas are near the kitchen.

> FOR MORE ON THIS HOUSE, SEE Case Study 3 (p. 109)

The accessible home gets the details right, like this wide sink with integrated drainboard, and an adjustable gooseneck faucet placed within easy reach.

Low windows connect the home with the natural world outdoors and also bring in the sun's warmth—important qualities for a homeowner who uses a wheelchair.

WHAT TYPES OF ACCESSIBILITY FEATURES ARE BEST?

Accessible, Universal, Adaptable, and Visitable: These words represent criteria used in designing for disabilities over the past half-century. *Accessible* generally refers to accommodations for people in wheelchairs. *Universal design* refers to places and objects that can be enjoyed by people of all abilities (see the sidebar on p. 12). *Adaptable* environments can be

SEVEN PRINCIPLES OF UNIVERSAL DESIGN

Universal design is a movement as well as an approach to designing products and environments that can be used by all people, regardless of abilities. Establishing the Center for Universal Design in 1997, architect Ron Mace articulated his Seven Principles to guide designers in creating objects and places that do not discriminate between people with disabilities and the able-bodied.

1. Equitable use: useful to people with diverse abilities

2. Flexibility in use: accommodates a wide range of preferences

3. Simple and intuitive use: easy to understand

4. Perceptible information: essential information, effectively communicated

5. Tolerance for error: minimizes hazards and the chance for user error

6. Low physical effort: usable with a minimum of fatigue

7. Size and space for approach and use: appropriate regardless of user's size, posture, or mobility

(*Source*: The Center for Universal Design at North Carolina State University)

The accessible home is adaptable. Here, the space above the dining area can be enclosed to add another bedroom in the future.

made accessible with a minimum of effort. And *Visitable* homes allow people with disabilities to enter and use the restroom. Taken together, these criteria provide the tools for tailoring the home to individual user needs.

This book puts the focus back on accessibility as a way to bring attention to the range of human differences and the need to create homes that are tailored to their occupants. In looking at a universally designed entrance it is easy to see the wide door and flat threshold as the whole story, whereas closer inspection will reveal many thoughtful details. Wind chimes alert the visitor with low vision that she has arrived at the right address. The homeowner with hearing loss has a doorbell with flashing lights to announce that a guest has arrived. Glass sidelights running the full height of the door let the homeowner seated in a wheelchair see who is on the porch. Space beside the entrance holds a seat where a visitor can rest while waiting for the door to be opened. These are features of the accessible home—subtle and powerful.

> FOR MORE ON THIS HOUSE, SEE Case Study 20 (p. 195)

The accessible home has great spaces for socializing, because friends' homes are seldom visitable.

THE POWER OF DESIGN

Well-planned accessible homes lift the spirits and enhance dignity, transforming our relationships with our bodies and our homes. Homeowners often report that renovations have changed the ways family members relate to each other. The pride of accomplishment shines through the eyes of a child with disabilities who can finally join in family chores. A responsive environment allows a mother with disabilities to manage everyday activities more easily, creating more time for her children. Children freed from caretaking roles can enjoy being siblings. Released from the need to provide constant assistance, a partner is freed to enjoy her disabled spouse's companionship. The reduction in tension is palpable when parents have a home that serves their family's needs. Accessible homes really do make everyone happier.

This book is about more than codes and gadgets, and it's about more than design. It's about vision and creativity, and the special alchemy that occurs when designers really listen and when clients really share their needs and hopes.

The accessible home has great places for being with the ones we love.

This book offers ways of envisioning how ordinary people with extraordinary challenges can partner with architects, designers, and their own families to create homes that restore capabilities, independence, and the grace of daily living. ✦

> FOR MORE ON THIS HOUSE, SEE **Case Study 21 (p. 199)**

The accessible entrance has ramps integrated into the forms and materials of the landscape.

DESIGN AS PROBLEM SOLVING

Design is creative problem solving. Equal parts art and science, design is the process of articulating a problem, exploring alternatives, and envisioning a solution, culminating in the creation of something new. Not to be confused with decoration—the embellishment of an object in order to create beauty—design is beautiful because it works, and it does so on many levels.

Design doesn't cost more. What is costly is not addressing the problem fully. When a new ramp doesn't complement the house's character and has to be removed before selling the property, the homeowner pays twice. When a bathroom has a great tile job but poor lighting, the owner's safety is in jeopardy. When homeowners muddle along with a dysfunctional kitchen, the cost of renovations needs to be weighed against the cost of nightly take-out. As you'll see in the case studies in the second half of this book, many designs actually save the homeowners money by increasing usable space without an addition.

CHAPTER 2

approach
and arrival

SEEN FROM THE OUTSIDE A home sends powerful messages, often subliminal, about who belongs. The accessible home takes control of that message: Everyone is welcome, regardless of abilities. That message is clear the moment the home comes into view. What does your home communicate to arriving visitors? Is there a level nonslip pathway to the entrance? Are steps sized for a comfortable gait? Are railings sturdy enough to lean on? Can the address be read from the street? Are walkways well-lit at night? These simple gestures show consideration for all who

Porches don't need to be large to be effective. A simple covered landing defines the entrance and provides weather protection as you enter this Maine home.

approach. Whether you are modifying an older home or designing a new one, this chapter will identify ways that make it easier to navigate the site.

PLAN CAREFULLY FOR AN ACCESSIBLE SITE

Accessible sites do not occur naturally, otherwise all environments would be user-friendly. Nature is hilly, bumpy, muddy, squishy, tangled, and uneven. Adding pavement brings a little order to the chaos, but unless the grading and drainage are carefully planned the pavement can be too steep to navigate easily or else too flat—an area for puddles that ice over in the cold. The accessible site requires planning, and the basic planning module is the footprint of a wheelchair, at 30 in. by 48 in.

Set the bar high in defining goals for site accessibility. The site is safer when there are separate areas for cars and people, with boundaries well marked to prevent meandering. It is easier to navigate when pathways are wide enough and slopes are gentle enough. Site design reduces stress when it includes clear

All elements of the path of travel—from sidewalk to porch to yard—are integrated into the site for safe and comfortable passage. Hedges and shrubbery define walkway edges, making it easy to stay on course. A landscaped "buffer zone" allows a visual connection with community life along with a sense of privacy.

SITE PLANNING
should start before
the home design
is set in concrete,
and include
circulation to and
around the house
for pedestrians, for
wheeled mobility
devices, and, often,
for automobiles.

sightlines and adequate lighting that help orient people as to where they are and where they are going. Ramps and stairs are more comfortable to use when handrails are properly placed. The site becomes an extension of the home when a person with a disability can pick up the mail, take out the trash, join friends and family in outdoor activities, and greet arriving guests. To accomplish these goals a project requires careful planning.

New single-family houses have wide latitude for creating an accessible site, and many areas now require at least an accessible "zero-step" entrance (see the sidebar below). Placing the main entrance near ground level ensures that ramps and stairs stay at manageable heights. Locating the garage close to the house keeps walkways short. Coordinating building floor-to-ceiling heights with the natural topography brings the outdoors within reach. Coordinating the design with sun angles and natural topography can reduce snow buildup and control rain runoff. All these techniques improve site safety and accessibility.

Although older houses pose many barriers to mobility, there are often ways to improve access. Design a network

of pathways linking outdoor activity centers. Indicate safe travel areas using contrasting walkway colors and textures. Construct ramps and gentle steps with sturdy handrails where the land slopes steeply. Add lighting for visibility. Reduce maintenance by providing proper drainage, durable paving, and sturdy native plant species. With accessibility as a goal, both the site and the house can be brought into greater harmony.

DESIGN FUNCTIONAL DRIVEWAYS AND PARKING AREAS

Start site design by identifying the personal mobility devices that will be used. These inform the choice of a van or automobile, which in turn sets requirements for driveway and parking areas. Some people with physical disabilities can transfer to a car seat, while others roll their chairs into specially adapted vans. Vans may be equipped with either a platform-type lift or an extension ramp. Van doors can be sliding, swinging, or upward acting, and located on either the side or rear of the vehicle. Once the vehicle requirements are known, the designer can then start to lay out driveways and parking areas.

DESIGNING FOR VISITABILITY

Eleanor Smith was traveling around Atlanta in 1986 when she noticed stairs on nearly all of the houses in a new subdivision, putting them off-limits to people like herself using wheelchairs. She founded the grassroots group Concrete Change to bring basic access to new houses, a concept she dubbed "universal visitability." Today many regions have enacted visitability laws, and standards for federally funded housing now require visitability. The visitable home is characterized by three important features:

- One zero-step entrance
- Doorways with 32-in. clear passage
- One wheelchair-accessible bathroom on the main floor

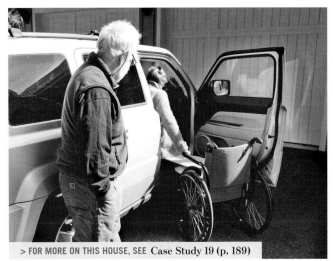

> FOR MORE ON THIS HOUSE, SEE Case Study 19 (p. 189)

When designing the garage, it's especially important to know how the homeowner transfers to and from cars, and what type of vehicle is used. Knowing whether the user requires assistance or can self-transfer makes it possible to design adequate space, and in the proper location. Here, Alice uses a transfer board to get in and out of her family's van.

Design the driveway layout and grading around the vehicle's turning radius and undercarriage clearance. Design maneuvering space to avoid long backups or extensive Y-turns. Distinguish driveway edges to prevent vehicles from driving off the paving, and mark pedestrian areas to reduce chances that people will wander into traffic, or that cars will encroach onto footpaths.

In designing the parking areas follow local zoning regulations for stall sizes, understanding that dimensions are minimums. A typical automobile stall is 8 ft. by 20 ft, and side-loading vans generally require an 11-ft. stall width. Measure the vehicle where possible, with all doors open and people actually entering, to see how large the parking area needs to be. Provide space for an accessible passenger loading and transfer zone, either beside or behind the vehicle, and leave an aisle on the driver's side for using the door. If transfer requires assistance, make sure there is space for additional people nearby. Mark pedestrian walkways and transfer areas within the parking area using a change in pavement texture or color for added safety.

WAY-FINDING

Way-finding refers to a system for organizing the physical environment so that people can better orient themselves in a place. Like color coding or signs in a public building, way-finding devices create a spatial language that we read with our senses.

Way-finding features can be incorporated into homes in ways that are subtle and effective. Changing the texture of walkway surfaces to differentiate edges or crosswalks promotes way-finding for people who rely on a white cane to find their way through touch. Raised curbs beside a walkway also allow way-finding when the homeowner is blind; placing the curb on one side only makes it clear which direction one is walking.

A change in paving color and texture clearly distinguishes pedestrian and vehicular traffic areas. By placing a curb to one side of the pathway, and using a light-color paving stone, this accessible walkway is easy to use by someone with low vision or a wheelchair.

SITE WALKWAYS in the accessible home should always be carefully detailed. Avoid creating walkways that will become displaced in the long term by removing tree roots nearby. Prevent erosion and movement with an effective drainage plan. Reduce damage done by frost-heaves by installing slab reinforcing and pouring thick slabs of paving material.

DESIGN AN ACCESSIBLE OUTDOOR PATH OF TRAVEL

Homeowners, visitors, and codes and standards all agree: Creating a suitable path of travel is the primary goal of accessible site design. An accessible path promotes independence by putting activity areas within reach. It reduces stress by making it easier for people to find their way (see the sidebar above). Elements of the outdoor path of travel include walkways, ramps, and site steps. Because we never know who will be a visitor or what the homeowners may experience after a home is built, the path of travel needs to be accessible to everyone:

■ For people with wheeled mobility devices, walkways should be planar and at least 3 ft. wide with turning areas at least 5 ft. wide.

■ For people with low vision, walkways should be texturally distinct and clear of obstructions that may cause injury, particularly around the arm and head.

■ For people who are blind and rely on memory to find the way, it is useful if walkways are straight rather than curved or zigzag, and if turns are at 90 degrees.

■ For those with hearing loss, walkways should have clear sightlines at corners and where vehicles are maneuvering.

■ For those with cognitive impairment, who may interpret a shadow or dark mark as a hole in the walking surface, even illumination and consistent coloring are essential.

■ For people walking, clear headroom should be at least 80 in. high.

■ For everyone, walkways need to be integrated into a site grading plan that controls the inevitable flow of water—away from the walkway surface.

SURFACES

Walkway surfaces should be level, uniform, nonslip, and slightly sloped for drainage. Whether paved with a poured finish or constructed of porous materials, the surface should be free of tripping hazards, which codes describe as level changes greater than ¼ in. Distinguish the walkway from surrounding surfaces by changing textures, colors, or materials. Pay special attention to walkway edges so that wheelchairs or navigational canes do not become stuck in the mud on wet days. This can be accomplished by bringing the lawn right up to the surface or by raising walkway edges. Consider installing a snow-melt system below the paving or providing covered walkways in areas where winters are harsh, to keep walkway surfaces clear.

RAMPS

Sloped walkways or ramps are preferable to stairs for an accessible path of travel, allowing a continuous walking surface from the street into the home. Designing a ramp requires balancing the site characteristics with the user's mobility needs; scooters, wheelchairs, prosthetics, and walkers require different kinds of adaptations. Short, steep ramps are difficult to use with manual wheelchairs, for example, and shallow ramps that

Wraparound porches with a low rail close to floor level put the outdoors within reach for an outdoorsman without sight. Changes in texture demarcate site walkways from porch areas.

> FOR MORE ON THIS HOUSE, SEE **Case Study 17 (p. 180)**

Using a combination of steps and ramps, this yard successfully and accessibly negotiates its slopped site.

By incorporating artistic guardrails, colored walls, and native plant materials, the architects have created a playful way to explore and move around this house's site.

Winding pathways weave through the accessible backyard. Ramp side-walls become transfer benches and put raised-bed planters within reach.

when designing pathways for safe daytime travel. During the evening, visibility is reduced by darkness and the presence of light fixtures that cause glare. Choose light fixtures for even illumination of walkway surfaces: step lights alongside walls, low mushroom-type fixtures nestled into the landscape, or up-lights placed at the base of trees. Locate light fixtures at decision points—beside the walkway or at directional turns, for example—to enhance way-finding. Control outdoor lights with motion sensors for a hands-free path of travel.

lengthen the walking distance are difficult for people with low stamina (see the sidebar on the facing page). Install raised curbs or low rails along ramp edges to prevent wheels from slipping off. Use trench drains rather than circular floor drains between flat and sloped paving to keep surfaces planar and reduce the chance of tipping during wheelchair turns. Weave ramps into the landscape by raising the surrounding grade levels and incorporating landscape plantings.

SITE STEPS

Since some people have difficulty using ramps, gentle site steps can be an essential part of an accessible path of travel. The most comfortable steps are compatible with a person's walking stride, typically 24 in. apart. Wider terraced platforms create places to rest and get one's bearings and simplify travel for those using walkers. As with walkways, use sturdy materials for both horizontal and vertical surfaces.

LIGHTING

Good site lighting makes it easier to stay on course and avoid falls. During daytime, bright sunlight casts shadows that may seem indistinguishable from holes in the pavement and can cause people with low vision or brain injuries to step off the pathway. Consider the natural shading qualities of plants and building elements

The goal of a lighting plan is to call attention to entrances and spill light on a path of travel. A continuous path from the parking area to the front door, flanked by low shrubbery, keeps the lines of sight open.

Step lights provide even illumination to make travel at night safer. Sturdy handrails mounted over the tops of posts give a continuous gripping surface.

DESIGN WELCOMING AND ACCESSIBLE ENTRANCES

As the transition between the outdoors and the home, the entrance has a lot going on. We want our guests to feel welcome and our family members to feel secure. Thoughtful amenities convey a message of caring, and accessibility features improve safety. Functionally, the entrance is a weather barrier, maintaining indoor climate control and keeping foul weather outside. Because the accessible entrance has outdoor walkways at the same level as indoor flooring, controlling the flow of water, mud, and snow can be a challenging task, whether the site is hilly or flat. Grading the site to drain water away from the house can be an important task in designing an accessible entrance.

THE LANDING

The accessible entrance starts with a landing—a flat surface where one "lands" after using a door, ramp, or stairway. Its primary function is to keep wheeled mobility devices from rolling backward or people from stepping off as the door is opening. Landings need to be large enough for people to reach door hardware and also to move aside as the door swings open. They also need to be configured so that people can position themselves to take hold of handrails before using the stairs or ramp back to grade. The size and location of the landing depends on the direction from which one approaches and whether the door swings inward or outward. Start with a minimum landing size of 5 ft. by 5 ft. so that a wheelchair has room to change direction, and plan the landing with space for additional people and gear to suit your own family's lifestyle.

GRADING BASICS

Site slopes are expressed as either a ratio or a percentage. A 1:20 slope means that for every 1 ft. in vertical distance (the "rise"), the horizontal distance is at least 20 ft. long (the "run"). As a percentage 1/20 = 5%. The following table lists common slopes under the ADA, listed from shallow to steep:

1:100 (1%)	Too flat in rainy areas, as water will not drain away
1:48 (2%)	Parking areas, cross slope of walking surfaces
1:20 (5%)	Walkways, do not require handrails. Steeper slopes are ramps and require handrails both sides.
1:16 (6%)	Comfortable ramps
1:12 (8%)	Maximum permissible ramp slope
1:10 (10%)	Short ramps, with a rise less than 6 in. (such as curbs and thresholds)
1:8 (12.5%)	Not allowed for an accessible site

Applying the 1:20 ratio to a 3-ft. hillside results in a 60-ft. walkway or a 36-ft. ramp. Since people get tired using ramps more than 30 ft. in length, most codes require switchbacks for longer ramps. Design level landings at least 5 ft. long at ramp top, bottom, and turns.

STAIRS

Unless the landing is at ground level, the entrance will need some kind of vertical travel device: either a ramp, or stairs and a lift, depending on the distance to grade. Stairway design is a matter of comfort and safety, available space, and code requirements, and an important item to get right. Stair widths should be at least 36 in., although the design of the house may suggest something wider. A shallow stairway, with deep treads (horizontals) and short risers (verticals), is generally easier to use than a steep one, although it requires a larger area and needs a wider yard. Many people find stairways comfortable when treads are at least 11 in. deep and risers are from 4 in. to 7 in. high. For safety, be sure tread and riser dimensions are uniform so that people do not need to alter their gaits, which can cause falls. Comfortable stairways are safer stairways.

This stair designed for a woman who lost a leg to illness omits tread nosings, because her prosthetic foot cannot straighten to avoid tripping. A platform lift at the porch landing offers an alternate way to enter the home with arms full of groceries.

RAILING SYSTEMS

Railing systems are an essential safety feature in stairway design and an important component of the accessible path of travel. Everyone occasionally needs something to hold onto, and handrails provide real safety as well as a sense of security. A railing system has two parts: the handrails that we hold onto and guards that enclose the space below to keep people from slipping off walkway edges.

Safety issues in railing design are established by codes, which set requirements for handrail heights (34 in. to 38 in. for adults and up to 28 in. for children) and limits on the size and shape of the gripping surface so that it can be comfortably used. Codes limit the size of guard openings to prevent entrapment, and also detail the requirements for handrail installation. Handrails should be continuous to make it easier to maintain a grip when floor configurations change, such as at landings. By extending the handrails past landings, people with low vision or assistive devices will have advance warning before arriving at any change in ground level. Railings should be sturdy enough to carry the weight of a hefty person (250 lb. minimum), and mounted securely to posts or walls. The accessible home has handrails on both sides of a ramp or stairway so that a person has support going both up and down.

LIFTS

Install a platform lift where the site lacks space for a comfortable ramp. A type of personal elevator with a flat floor rather than an enclosed cab, this lift is designed to ride vertically alongside a short mechanical tower, for distances up to 12 ft. Low walls and curbs keep wheels from rolling off, and gates can be placed for either roll-through use or 90-degree turns. Subject to access and elevator codes, lifts may require enclosures or special controls, so check requirements in your area. Lift equipment offers choices regarding platform size and controls, so it is useful to visit a showroom or test-drive an installation to find the right lift for the site.

FIND A WAY to test-drive various stairways to find your own comfort levels; check out friends' entrances or have the builder create a mockup, paying attention to railing systems at the same time.

A homeowner using a power wheelchair would not find a handrail useful. A guardrail, however, is an important safety feature here, preventing falls to the basement patio below, with small openings to prevent entrapment.

A platform-type chair lift makes a porch accessible when the site does not have space for a long ramp. One enters the platform at either landing and rides up or down.

> FOR MORE ON THIS HOUSE, SEE Case Study 21 (p. 199)

> FOR MORE ON THIS HOUSE, SEE Case Study 2 (p. 103)

A simple entrance carved into a basement wall has all that's needed: a wide door with sidelights, non-glare lighting, a level floor, and a mailbox within easy reach from a seated position.

A simple design gesture—extending the roof over the porch—combines with a recessed doorway and warm lighting to make a welcoming entrance. Roof gutters and downspouts drain water away from the path of travel from driveway to entrance.

COVERED LANDINGS

Traditional homes in many parts of the country may or may not have covered landings, but in the accessible home a roof has an important function. With the landing at floor level and the threshold installed flush, rain and melting snow will tend to seep under the entrance doorway. A landing roof provides weather protection, and larger roof areas provide more protection than small areas. Depending on the climate, an accessible home will have a porch with a roof.

OTHER AMENITIES

Thoughtful entrance amenities appeal to visitors of all abilities. A porch light is more than a convenience, it's a safety feature and eases the transition between darkness and light while entering the home. Consider providing a place for cleaning mud off mobility devices, with a nearby faucet or suitable decking material to remove grit from shoe and wheel treads. Choose a doorbell for visibility from the exterior and audible (chimes, buzzers, bells) and visual (lights) signals

on the interior. Locate the doorbell at an accessible height and with clear space for a wheelchair beside it. Provide an oversize mailbox so that packages do not clutter the landing. Add a shelf for placing bags while rummaging for keys, and a bench where a person can rest while waiting for the door to be opened. The house number should be large enough to be legible from the public way, and placed where it can be easily seen by day or night. Integrate these features into the design for an entrance that is attractive, functional, and secure.

Now that we have arrived at the front entrance, it's time to step inside. That same attention to the safety and comfort of residents and visitors that characterizes the accessible site also applies to the home. Let's see how it all works. ✦

A low doorbell allows this homeowner with a wheelchair and limited dexterity to call for assistance when she returns home.

CHAPTER 3

building connections

ASK MOST PEOPLE WHAT MAKES a home wonderful and you'll hear about the rooms—a great kitchen, a fun family room, or a luxurious bathroom. For the accessible home, it is also about the spaces that lie between the rooms. What makes a home wonderful is the ease of moving between spaces, from outdoors to indoors, from room to room, from one floor to another.

Like its outdoor counterpart, the site, the interior of the home needs an accessible path of travel that makes it easy to get from one place to another. Designers call this the circulation system, and, like

The architect was able to connect three floor levels within a 9-ft. vertical rise by installing an enclosed platform lift in a small space between the original house exterior wall (with artwork at left) and the property setback (outdoor stair at right). Skylights and glass doors help make closed quarters seem spacious.

> *"When we design something that can be used by those with disabilities, we often make it better for everyone."*
> —DONALD NORMAN, SCIENTIST

> FOR MORE ON THIS HOUSE, SEE Case Study 19 (p. 189)

the veins and arteries that link the body's organs, these pathways link functional areas. We've discussed the way outdoor circulation links spaces for cars and people, adapts to flat and hilly sites, and connects the yard with the entrance. Inside, the house's circulation system links entrances (connecting indoors and out), halls and doors (connecting rooms), and stairs and elevators (connecting floor levels). Let's look at how each of these connections can be made more accessible.

CONNECT OUTDOORS WITH INDOORS USING ACCESSIBLE ENTRANCES

The accessible home makes it easy to enter and exit, and all exterior doorways—the front door, back door, patio and porch doors—serve dual functions. As entrances they are generous and welcoming. As exits they are never too far away when we need to get out. Doorways are wide enough for people and their gear: groceries and luggage coming in, food for the barbecue and sports equipment going out. Coming

Wide doorways and level floors offer safe passage between rooms.

Transom glass and side lights bring abundant natural light into the entrance, and allow a homeowner to see arriving visitors. Combined with a wide door and zero-step entrance, these features make a welcoming and accessible point of arrival.

COORDINATE SCREEN DOORS with the entrance door. A retractable screen is an effective complement to a swinging door because it slides out of the way when not in use. Sliding motions are easier for many than the push-pull action of a swinging door.

in, the door can be unlocked and the handle manipulated easily. Going out, especially in an emergency, requires minimal effort, a simple push. As entrances, doors put the safety and comfort of home within reach. As exits, doors put nature's freshness and the vitality of the neighborhood within reach. Doors keep out the insects and keep in summer's cool and winter's warmth. It seems like a tall order for an entrance to accomplish all this, but the requirements are surprisingly simple.

DOOR SELECTION

With so many attractive designs to choose from, it is easy to think that the choice of doors is primarily a stylistic decision, but there are functional issues involved. Panel-type door surfaces should be smooth for the lowest 10 in., so that wheelchair footrests do not get caught in decorative moldings. Vision lights (windows either in or beside the door) should enable someone seated to see outside, with glass not more than 43 in. above the floor. As exterior doors are part of the "building envelope," the entrance must also comply with fire safety and energy codes, so check local

requirements before finalizing a purchase. Door thresholds in the accessible home are flush with the floor, or up to ½ in. high (maximum) and with beveled edges.

HARDWARE

Entrance hardware should be comfortable to the touch and intuitive to operate, usable with a closed fist rather than a gripping or twisting motion. The backset (the distance from hardware centerline to door edge) should be wide enough to prevent hand injury. Choose lever hardware rather than knobs, mounted 34 in. to 48 in. above the floor. Hardware should allow the door to be operated without too much effort (5-lb. pressure or less). Egress hardware should include closers that prevent doors from shutting too quickly (generally at least 5 seconds for an 80-degree arc). Choose keyless egress hardware for quick and easy exit in an emergency. Door hardware also includes hinges, thresholds, and stops (wall, floor, or overhead models) that keep the door from pulling out the hinges. Wherever possible, "test-drive" door hardware before purchasing, to be sure you can enter and exit the home easily.

LOCKSETS

There are good alternatives to standard keys if we look to the security industry. Electronic key-pads that respond to a user access code, proximity card-readers that respond to a swipe of an ID card, and fob-activated locks are three options. Biometric security devices recognize fingerprints, faces, and the eyes of pre-approved users, offering an alternative to passwords for security controls. Magnetic locks and electric strikes can be used for "buzz-in" systems. Automatic door operators rely on push-button controls, similar to elevator doors. With a little research you can find locking hardware that is just right for you.

> FOR MORE ON THIS HOUSE, SEE **Case Study 23 (p. 213)**

Glass doors bring daylight into the house and allow someone with hearing loss to see who is in the next room without entering. Wide hallways make it easier to get around using mobility devices or with assistance. Color contrast between floors, baseboard trim, and walls improves visibility for those with low vision.

REMOVE BARRIERS TO CONNECT INDOOR ACTIVITY AREAS

To make the rooms in standard houses larger, the building industry has long been making hallways and doors smaller. Standard interior door sizes are 30 in. in most homes, which can translate to a $27^1/_2$-in. opening measured between the face of the door and the jamb stop in the open position. Interior doorways actually need to be at least 32 in. clear for a wheelchair, so doors should be at least 34 in. wide. It is just as easy to install a wider door as a small one during construction—a little less drywall and framing, a little more finish trim, a fair trade-off.

Smaller doors allow smaller hallways, and many houses have 34-in. hallways. For an accessible home, clear hallway width should really be 36 in., measured from the wall surfaces. When wainscots and chair rails narrow the clearance between walls, then the hallway should be widened. Rough-framing hallways at 37 in. wide is also an easy task during construction, but in renovation the work is more complicated and costly. Wall switches

Windows and doors alternate with built-in shelving to reduce the sense of distance along an interior walkway. Window seats at flat landing areas provide a place to stop and enjoy the view or read a book.

represent barriers
to accessibility
when they do not
allow people with
mobility devices to
navigate through
a building. Seeing
these basic building
blocks as barriers
allows designers to
start to envision
a different kind
of home. It's a
necessary and
transformational
step in creating an
accessible home.

need to be relocated, floors and ceilings
patched, perhaps structural modifications
made. There are better options, and
the designer needs to question basic
assumptions about the home to find them.

How wide do the halls really need to
be? When one person in a family has a
wheelchair, chances are good that friends
and visitors will also have mobility
devices. Many accessible homes have
hallways that are 60 in. wide, roomy
enough for two people using wheelchairs
to pass each other. It's also highly likely
that the crowd will gather at the home of
the person who uses a wheelchair because
other homes are not visitable, and so
hallways need to be wide enough for
gatherings. Wide hallways are only part of
the answer.

When we eliminate interior walls,
interesting things happen. Rooms become
wider, dead-end passageways disappear,
and noise levels change (see the sidebar
on p. 39). Sightlines are formed between
activity areas, which is a good thing for a
parent who previously had to walk into the
next room to check on the kids. The house
is bright and airy, with sunlight streaming
into interior spaces and cross-ventilation
between rooms. For people with hearing
loss whose mode of communication is

visual (lipreading, sign language), those in
the next room seem closer. For people with
low vision whose mode of communication
is auditory, sounds in the next room are
easier to hear. Many accessible homes
have interior spaces that are like lofts,
with furniture groupings at the dining and
living areas separated by aisles rather than
by walls.

DESIGN HORIZONTAL CIRCULATION FOR INDEPENDENCE

An accessible home supports people's
independence by creating opportunities
for people of all abilities to get around
safely. Start by identifying the requirements of those who will be in your home.
There is much that can be done to modify
an older home to improve the path of
travel, and even more if you're building
a new home. Small improvements have a
big impact.

DESIGN FOR MOBILITY DEVICES

Wide circulation routes are necessary but
not enough for safe horizontal travel using
mobility devices. The accessible home
has flooring, wall treatments, and door
placement all designed to promote ease of
travel and durability of surfaces:

The stairway and
adjacent elevator
are on the main
indoor path of
travel. Generous
windows beside and
at the end of the
hallway ensure both
safe passage and
orientation within
the house.

- Locate doors to enable wheeled mobility devices to turn without backing up.
- Angle the walls at corners to widen the space where it is most needed.
- Align doors across a hall to shorten the path of travel between rooms.
- Use wood wainscoting to protect plaster from being nicked by assistive devices.
- Avoid gaps in wall surfaces and railings that might catch assistive devices or clothing.
- Install 9-in.-high baseboard moldings to protect wall surfaces from damage by wheelchair footrests.
- Adjust underlayment thickness to accommodate different surface materials so floor levels do not vary by more than $1/2$ in. Bevel edges to keep vertical surfaces $1/4$ in. or less.
- Securely attach carpets and backing.
- Select in-floor heating registers so openings are no more than $1/2$ in. wide and perpendicular to the path of travel.
- Lay out furniture arrangements to gain space for travel within a room.

DESIGN FOR AGING IN PLACE

We can't control the aging process but by making homes that help prevent accidents, we can reduce the chances of hospitalization. Arthritis, back problems, and heart and respiratory diseases are a normal result of aging, and make people more prone to injury. The accessible home makes it easier to avoid falls by keeping the walking surfaces clear and safe. Design level sturdy floors, free of obstructions that could cause a person to slip or trip. Install carpets flush with the floor. Use skid-resistant and glare-resistant finishes such as textured tile or satin-finish urethane on wood flooring. Design adequate storage at stair landings so that items are not stacked on stair treads.

TRAILING

People without sight find their way using touch in a process known as *trailing*. Standing with one arm about 6 in. from the wall and the hand extended about 12 in. from the body, a person walks forward, keeping the back of the hand in contact with the wall and the fingers slightly cupped. When an object is encountered, the person takes a few moments to examine and identify it through touch. The designer should ensure that interior circulation areas are free of items that could cause injury as people bend at the knees to explore objects in the environment. Call for easy-clean wall surfaces and decorative moldings as well as flooring changes to mark an accessible indoor path of travel. Wood paneling and chair rails become trailing devices when the height is coordinated with the user's reach range. Coordinate trailing features with electrical switches and outlets and plumbing installations to make controls easier to find using touch.

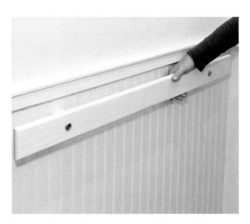

A wood chair rail allows people with low vision to navigate within a room using touch, and a beadboard wainscot makes a durable wall in areas of heavy use. A custom-designed grab-bar, strategically placed, increases safety at walkways.

DESIGN FOR LOW VISION

A home where it is easier to see one's way is a safer home. Place electrical outlets at the points of use to avoid having loose electrical cords that create a tripping hazard. Provide adequate lighting to keep walkways visible. Keep wall surfaces free of protruding objects that could cause a person to become injured. Design concepts

LANDMARKS

Landmarks are built-in features that mark destination points or that trigger decisions such as a change in direction. The designer can use landmarks to create homes where it is easier to navigate with low vision. For example, the hum of a ceiling fan is a landmark that helps people hear their way to the family room. Flooring that differentiates between rooms and textured thresholds are also landmarks. Bathroom floor tiles can be varied to indicate where it is advisable to stand to use the plumbing fixtures.

used in environments for people with low vision, such as "landmarks" and "cognitive mapping," provide insights for creating homes that improve visual accessibility (see sidebars above and on p. 29).

DESIGN FOR INTELLECTUAL DISABILITIES

Spatial disorientation associated with dementia can lead to anxiety, fear, and safety problems such as getting lost. Clearly mark critical areas such as the bathroom, and camouflage areas that are off-limits, such as a utility closet, to make a home that discourages unsafe wandering. Contrast walls with doorways to distinguish room openings. Install baseboard moldings, walls, and floors in contrasting colors to clearly differentiate down from up, preventing confusion. A person with dementia may read strong flooring contrasts as holes and refuse to pass by, so use uniform colors and avoid shadows and glare (see the sidebar on p. 32).

INCORPORATE "GADGETS"

As part of the home's egress system, the hallway is a natural place for various alarm systems, including fire and smoke detection and security devices. Select doorbells, alarms, and security systems for users who have low vision (with audible signals such as bells and buzzers) and/or

hearing loss (with visible signals such as flashing lights). In most homes, electrical controls are in the hallways—lights to and from adjoining rooms. Consider specifying occupancy sensors so that light fixtures turn on when a person enters the hallway. Planning for these devices within the design is the best way to ensure they are attractive and functional.

TAILOR DOORS AND HARDWARE TO ROOM FUNCTIONS

Hinged swinging doors can be difficult for people in wheelchairs, who need to roll aside as the door swings open. Closing the door is equally challenging, and may require rolling into the next room to reach the handle, then backing up while pulling it shut. Accessible doorways need space for a wheelchair beside the latch edge, from 12 in. to 48 in. wide, depending on the angle of approach and the direction the door swings. Replace standard hinges with offset hinges to move the door out of the framed opening as it opens. Many of the homes in this book have swinging doors for bedrooms, as they close tightly and can be locked.

"Use lever handles and cup pulls rather than knobs for hardware. They are easier to grasp, and can be used by someone with weak hand strength."

—JOHN SALMEN, ARCHITECT

SLIDING DOORS

Sliding doors are a good option where a swinging door would block a path of travel, and many people find them easier to operate. New hardware options make sliding doors a good choice for room entrances. Sliding on overhead tracks

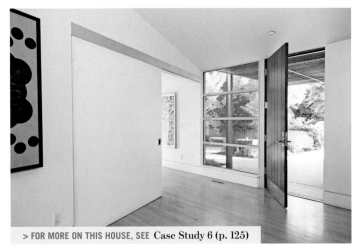

> FOR MORE ON THIS HOUSE, SEE Case Study 6 (p. 125)

This 5-ft.-wide sliding door separates the bedroom wing at left from public areas of the house, for both privacy and sound control. By recessing the wall surface under the overhead track, the door lies flush against the wall in the open position, leaving maximum space for furniture placement and circulation.

mounted on either the ceiling or wall, doors can be made to stack, such as room dividers, or to slide on the face of a wall, such as a barn door. Access codes limit sliding door weight, so that doors are operable without undue physical pressure (an adjustment that is made in the manufacture of hardware). Many of the homes in this book use sliding doors at wide hallways that separate sleeping areas from living areas.

POCKET DOORS

For rooms made too small by a swinging door or one that is left open much of the time, doors that slide into a pocket in the wall are a popular option. Pocket doors hang from a track above the opening, so they do not close as tightly as swinging doors. Size the opening with hardware in mind, as standard pocket door hardware (edge pulls) can be difficult to use. If you are changing over to pocket doors in a renovation, in-wall electrical and plumbing fixtures may need to be moved. Many of the homes in this book use pocket doors for bathrooms.

BI-FOLD DOORS

Bi-fold doors are narrow panels connected by hinges. Each panel is half the size of a standard door, making bi-folds a good option for closets, as they intrude less into circulation space. In the open position, bi-folds will reduce the doorway opening width unless hinges are selected that allow doors to fold flat against each other. Many of the homes in this book use bi-fold doors for bedroom and utility closets.

Bi-fold doors are a good choice for pantry and utility closets because they keep the path of travel open. Glass doors are used for this home's exterior entrances.

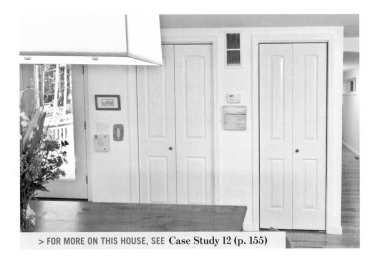

> FOR MORE ON THIS HOUSE, SEE Case Study 12 (p. 155)

COGNITIVE MAPS

Teasing apart the steps involved in memory allows the designer to create places where people with brain damage can function more effectively. It also enables us to design places where people who rely on memory, such as those who are blind, can navigate easily. Using our senses, we read the environment as a series of "navigational cues." Tile tells us we are in the mudroom, while hardwood floors say hallway. Our brains assemble these cues into "cognitive maps," a kind of spatial memory bank that we reference in getting around a place. Our internal maps allow us to find the bathroom in the night—counting the steps, listening to sounds all around. They also make it possible for a person without sight to function with ease in a familiar setting.

Lessons learned in therapeutic environments for people with brain damage can contribute to the design of homes for those with intellectual limitations. Dementia affects the mind's ability to form and use cognitive maps, resulting in confusion, anxiety, and fear in an unfamiliar setting. Whether or not a person actually remembers the way, navigational cues prompt them to pay attention. By "layering" spaces—for example, alternating between small and large spaces, high vs. low ceilings, active vs. silent areas, hard vs. resilient flooring, textured vs. smooth surfaces—the designer activates the memory to differentiate areas of the home. The fully accessible home is designed as a rich, multisensory experience.

DESIGN VERTICAL TRAVEL FOR SAFETY

Older homes built for active and fit young families, with stairs linking floor levels, do not reflect today's realities. In the past, life spans were shorter and many people were institutionalized, but today four out of five people want to live out their years at home. What can be done to make an older house more accessible? And when we plan new homes for multistory living, how can we ensure they can house us when our needs change? There is much that can be done to make homes with multiple floor levels more accessible.

STAIRWAYS

Stairways will always be a fact of life, a compact and inexpensive way to travel between floors. They are, however, a potential falling hazard, and thoughtful design can make them safer. Dimensions and details can be manipulated to make stairways easier to use. Sometimes a stair can be rebuilt to gain extra space—for example, stretching out a steep service stairway to an actively used basement. Principles of outdoor stairway design apply equally to interior stairs (see chapter 2, p. 14), although generally the latter are slightly steeper. Existing houses with limited space can still benefit from modifications that make a stair easy to use:

- Use nonslip finishes.
- Mark tread edges using a contrasting material to make it easier to see the drop off while descending.

A rosewood strip inlaid into treads make it easy to see changes in floor level, whether walking up or down the stairs.

- Contrast risers with tread colors to make it easier to see while ascending.
- Mount handrails firmly on both sides so that a person with one strong arm can get a firm grip going in both directions. Contrast railing colors with the walls for better visibility.
- Design handrails to be continuous or to extend past the landings as tactile and visual cues that the stairs are nearby.

RAMPS

Ramps are an excellent alternative to stairs for short vertical distances, such as connecting a step-down mudroom with living areas. A 6-in. step requires a 3-ft. ramp for comfortable wheeled travel at a 1:12 slope. As with stairs, landings and handrails improve safety. Mark the walkway surface clearly to announce changes in floor planes to prevent falls triggered by a shift in gait.

Short ramps can be added in place of single steps between floor surfaces, such as between this mudroom and living area. Glass doors and a low wall beside the ramp improve safety by keeping sight lines open.

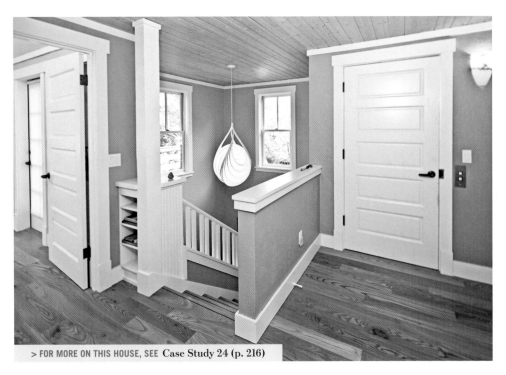

> FOR MORE ON THIS HOUSE, SEE Case Study 24 (p. 216)

By locating the elevator (door at right) beside the stair, vertical travel is consolidated and made equitable, with everyone—both abled and disabled—sharing the same routes. Shelves at each landing allow the homeowner to consolidate trips between floors, offering a place to store items that need to be carried up or down.

> FOR MORE ON THIS HOUSE, SEE **Case Study 22 (p. 207)**

This residential elevator is much used for carrying groceries to living spaces on the second floor. Continuing the wall and floor treatments from the landing into the cab makes the elevator an integrated part of the home.

ELEVATORS

Residential elevators and lifts are good alternatives to stairs in places where ramps are impractical. One of the biggest investments in creating an accessible home, an elevator well-integrated into the design can add value at resale. The choice of an elevator depends on many factors, with vertical travel distance, floor plans, and costs topping the list. There are four basic options:

- Residential elevator: The cab rides up and down within an enclosed shaft of fire-rated construction, extending beyond the upper and lower floor levels of a multistory building.
- Vertical platform lift: An open platform with low sidewalls rides for short distances, 12 ft. or less, and is less costly than an elevator.
- Incline platform lift: A flat platform rides beside a stair on a wall-mounted track. Because it folds against the wall when not in use, it needs a wider stair.
- Incline chair lift: A seat mounted on a wall-mounted track rides alongside a stair.

The building layout determines whether you can use a roll-through model or one that requires turns or backing out. Place the elevator near an interior stairway to economize on landing space and to allow everyone to use the same path of travel. If the elevator is added to an outside wall, the rooms beside it become landings and the house loses windows. Wherever the elevator is placed, it will have an impact on both horizontal and vertical circulation, so careful planning is important.

There are many choices to be made, and a visit to an elevator sales office for a test drive is advised. Functional mechanisms include pneumatic and overhead cable, each with advantages and drawbacks concerning noise, maintenance,

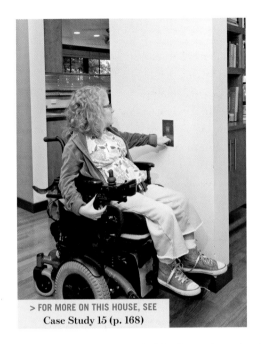

> FOR MORE ON THIS HOUSE, SEE
> **Case Study 15 (p. 168)**

ELEVATORS AND THE CODE. Several homes in this book have added windows, skylights, or transparent doors to keep small elevators from feeling claustrophobic, but as state codes may limit the amount of glass permissible, check local regulations. Lift requirements may be more lenient.

Elevator call buttons mounted within the user's reach and a short distance from the hall door ensure that she is out of range of the door swing while it opens. At the same time, the automatic door operator closing speed is adjusted to give her time to enter and leave the elevator safely.

and cost. Finish options include surface materials and lighting. Control options include push-button calling or voice activation, and audible or visual signals when the cab arrives at a floor.

We have seen how responsive design can eliminate obstacles to getting about, both inside and out, and make homes more user-friendly for residents and visitors alike. How can the same approach to design affect activities of daily living? Let's start with the rooms where we spend much of our time and invite our guests—spaces typically called living and dining areas. ✦

> FOR MORE ON THIS HOUSE, SEE **Case Study 9 (p. 140)**

A translucent plexiglass elevator cab makes vertical travel more enjoyable and helps orient the rider within the house. By placing the elevator within the stairway, routes of vertical travel are shared by all users.

CHAPTER 4

living and dining

MULTIFUNCTIONAL FAMILY rooms have replaced single-purpose living and dining rooms as the focus of family life. Passive solar designs capture sunlight and prevailing breezes to control indoor climates naturally. Open floor plans are the physical expression of these trends, and also the hallmarks of the accessible home. All that is needed is to ensure that the activity centers are themselves accessible. Let's take a look at shared spaces for gathering and recreation: the home's living and dining areas.

A small home feels larger when rooms are connected visually. This dining room in a converted barn links to the kitchen by a low wall and with the living room by a shared cathedral ceiling.

"The world is not what is handed to you, but rather, what you make of it."

— BRIAN MCMILLAN, HOMEOWNER

DESIGN FOR HOW YOU LIVE

Yesterday's living rooms were formal adult settings used for entertaining, quiet pursuits, and special occasions. Far away in the basement were the recreation rooms, boisterous places for children's active play. Great rooms represented an evolutionary giant step in the late 20th century—spacious rooms where both adults and children gathered for recreation and relaxation. Over a family's lifecycle the great room morphed from toddler playroom to teenage game room, then to crafts workshop and home theater for empty-nesters. Yet each of these functions has different requirements, with the result being not-so-great rooms, unsuitable for many activities, alongside lifeless living and dining rooms.

The accessible home takes a different approach. Instead of planning a great room, think about the kinds of places that make you happy. To design a home that is right for you, start by making an inventory of how you actually like to live.

- What do you enjoy most about your home's living spaces?
- What are you unable to do at home, but would like to?
- How do you like to spend time alone? With others? Seasonally?
- What kind of adaptations do you need for these activities to take place?
- How do you expect your lifestyle to change in future years or decades?

Seating clusters surrounded by wide aisles are user-friendly for those passing through the room or engaged in a group activity. Space for a wheelchair is designed into the layout.

> FOR MORE ON THIS HOUSE, SEE Case Study 19 (p. 189)

When we disregard ideas of what living and dining spaces need to look like, we can start to create places that really fit our needs. You may discover you need less rather than more space—or different kinds of spaces. Since the accessible home is tailored to the homeowner's abilities, activity centers will vary for each family. For example, if you entertain rarely but enjoy jewelry-making, consider devoting part of the living room to a workbench. Design walls with wide pocket doors to adapt smaller spaces for larger groups. Home at its most basic level is a place where you can be yourself, so pick and choose among the following options to create a variety of places for enjoying life.

DESIGN COMFORTABLE PLACES FOR BEING WITH OTHERS

Home is not just where we live, it is also where we host gatherings with friends and family. One family featured later in this book sized their new home with space for multiple wheelchairs—for their daughter and her friends (see Case Study 15, A Child-Centered Home, p. 168). A woman who found it difficult to travel created inviting spaces for hosting others (see Case Study 7, Accessible Farmhouse and Barn, p. 131). When friends' homes are not accessible, having great places to socialize is less luxury and more necessity. Visitable homes have reached the mainstream, with many communities requiring that new houses have accessible entrances and interiors (see the sidebar on p. 16).

LIVING

Whether you prefer to cuddle up with a book or entertain large groups, your living spaces should match your lifestyle. For one-on-one activities, design window seats and alcoves along the path of travel. For larger gatherings, lay out furniture with wide passageways for people with mobility devices. Link activity centers by ensuring clear sightlines that make

> FOR MORE ON THIS HOUSE, SEE Case Study 21 (p. 199)

ARCHITECTURAL ACOUSTICS

Voices in the air, footfalls on the floor—sounds are initiated when a stimulus creates vibrations in matter. Sound energy travels out from the source in all directions as waves, like ripples on a pond when we toss a stone. Some waves are absorbed by soft surfaces (recall the quiet of a wall lined with books). Some are deflected off hard surfaces (this is the backup chorus when we sing in the shower). Some waves pass through materials (an overheard conversation in the next room). And some sound waves eventually reach the ear, where tiny bones and membranes transform sound waves into fluid waves, and then to electrical impulses to the brain. The brain interprets these impulses in the process called hearing.

Architectural acoustics is the management of sound within a space. By knowing how different materials affect sound transmission, and also optimum sound levels for different activities, the acoustic engineer creates rooms that make it easier to hear. Sound is described in two ways: loudness and pitch. Loudness, or amplitude, is measured in decibels (dBA), with higher decibels meaning louder sounds. (At close ranges human speech measures between 50 and 60 dBA, and a jet engine is 140 dBA.) Pitch, or frequency, is measured in Hertz (Hz), and lower Hz ratings mean deeper sounds. (Most speech falls in the 250 to 4,000 Hz range.) The sounds all around us represent a wide range along each spectrum.

The acoustic engineer measures sound by initiating a noise and timing the number of seconds it takes for the sound to decrease by 60 dBA. For normal speech a reverberation time of 1 second (RT1) is optimum, whereas choral music benefits from the blending that occurs with RT2. Larger rooms and cathedral ceilings have a higher RT, as random sounds bounce off surfaces before reaching the ear. RT can be lowered by using building products with a high Sound Transmission Class (STC) rating—that is, a greater capacity to block sound. There are many ways to reduce sound in a space, and simple modifications have a big impact.

- **Building structure:** De-couple floor/ceiling assemblies and install solid bridging between joists to reduce vibration. Isolate joists from sheathing and strapping. Install vibration damping between underlayments.
- **Room configuration:** Design tray or coffered ceilings to absorb extraneous sound. Avoid parallel surfaces;

A coffered ceiling over the dining room table absorbs extraneous noise and focuses sound back town to the table, making it easier to hear.

skew walls slightly to eliminate echoes. Right-size rooms and provide wide openings.

- **Wall design:** Increase mass and air space. For example, a standard wood-stud wall with ⅝-in. drywall on each side has an STC of 35, but doubling-up the drywall, staggering studs, and installing the drywall on resilient clips increases the STC to over 60.
- **Windows and doors:** Choose windows with double or triple layers of glass and acoustic glazing compounds to block outdoor noise. Weatherstrip exterior doors and choose solid interior doors.
- **Finishes and furnishings:** Install soft surfaces such as carpeting and pads, cork flooring and underlayment, acoustic wall panels, and fissured acoustic tile ceilings. Choose upholstered furnishings and fabric window coverings.
- **Built-in cabinetry:** Specify "soft-close" door hardware to avoid slamming.
- **Electrical:** Avoid cutting holes in acoustic walls. Surface-mount electrical devices and fill openings with acoustic sealant.
- **Mechanical:** Locate heating/ventilating/cooling equipment away from living areas. Install using resilient curbs or suspension systems. Insist on round ductwork.

The field of technological sound enhancement has made exciting breakthroughs to enhance hearing. Many devices are portable and require simply an electrical connection, but others must be wired in during construction. Audio induction loops are cables installed around the perimeter of a room or building to transmit speech directly into hearing aids, without the distraction of random environmental sounds. Usually associated with places of assembly, looping technology is making its way into the home.

it easier for people with low vision or hearing to connect. Provide generous lighting to make it easier to see gestures and expressions. Improve visual communication by planning circular furniture layouts with chairs rather than sofas, and using wing-back chairs that amplify sound much like cupping the ears does. Larger rooms are inherently noisy, and benefit from acoustic features that block unwanted noise and enhance desired sound (see the sidebar on p. 39).

"The main floor is on the open plan, which makes it easier to hear the phone or one of its extensions ringing."

—HARD-OF-HEARING HOMEOWNER

DINING

Regardless of where eating takes place, the dining area is a center of family life. Whether located in an expanded kitchen or a corner of the great room, the family table is multifunctional, pulled into service as a homework center and cleared off at dinner time. In creating a dining

area, balance requirements for people and furniture. Use round or oval tables to allow those who rely on lip reading or American Sign Language to "hear" the conversation visually. Provide acoustic treatments, as too much sound reverberation degrades one's ability to participate in group discussions. For people with mobility impairment, size the room for furniture plus assistive devices. Choose tables with open knee-space for people in wheelchairs—a pedestal or trestle base rather than four corner legs. Select window coverings that filter daylight to avoid glare (see the sidebar on the facing page) and absorb sound.

IN THE DETAILS

Materials and details should always be selected to make a home comfortable and attractive, but in the accessible home, these decisions are more than aesthetic— they also make a home livable. Common sense will take you far in choosing products. The accessible home is low-effort, so use durable scratch-resistant wall materials to reduce maintenance. Place windows sills low enough to see outside from a seated position, and specify easy-to-use hardware such as cranks or remote controls. Consider adding an enclosed direct-vent gas fireplace with electronic ignition for warmth and ambiance without the worry of sparks flying or the work of dealing with firewood and ash.

CREATE STURDY PLACES FOR ACTIVE RECREATION

An explosion in television capabilities over recent years has changed the way designers think about media rooms: more as theaters, and less as living rooms. Media room size is a function of the type of equipment installed, optimum viewing distance, and the number of

Built-in cabinets put table linens and serving supplies within easy reach at the dining area, and also support serving counters that visually separate the two spaces.

occupants. If the gang gathers at your place for Sunday football, you'll need a different kind of media space than a movie buff who watches films into the wee hours. Consider adding a kitchenette for snacking. Acoustic, lighting, and window controls are essential to block extraneous sound and glare and make it easier to watch the screen. For the accessible media room, leave space for a wheelchair, place electronic controls within reach, and use remotes for all things mechanical (window blinds, ceiling fans, and television).

Comfortable chairs, clear sightlines, and a large-screen television make a basement media room the best place in town for watching sports.

LIGHTING THE HOME

Adequate light promotes safety, prevents accidents, reduces eye strain, helps people orient themselves in a room, increases one's sense of security, makes it easier to perform large and small visual tasks, and reduces depression. For people with vision or hearing loss, a well-lit room makes it possible to follow a speaker. Getting the light right is a top priority in designing the accessible home. A standard measure of light is the foot-candle (fc): the light cast by a standard candle placed 1 ft. away from a white surface. A sunny window reads approximately 1,500 fc on a light meter, and most living areas range from 15 to 100 fc according to various activities. Close work with small objects of little contrast over a prolonged period of time benefits from even higher light levels, so it's best to have professional assistance in tailoring the lighting to home activity centers.

The central ceiling fixtures found in many older homes put corners and work surfaces in shadow, and complicate the task of replacing burnt-out bulbs for a person with disabilities. Therefore, the accessible home has a variety of light fixtures, selected for layering of light, performance, and ease of use. Following are goals of an effective lighting plan:

Ambient lighting (10-20 fc) offers general room illumination. To keep pathways free of electrical cords, use decorative wall sconces or ceiling fixtures. Consider cove lights—fixtures mounted behind a valence that bounce light off the ceiling while hiding bulbs from view.

Task lighting (30-50 fc) is used for work areas such as a desk or kitchen counter. Use pendants over tables and under-cabinet lighting at counters.

Accent lighting (at least 30 fc, or three times the ambient light level) focuses on points of interests, such as artwork or plants. Picture lights and track lights are common options.

Egress lighting (10 fc for safe passage) illuminates the path of travel on stairs and hallways. Use step lights or low levels of ambient lighting.

Emergency lighting provides safe passage in the event of a power outage or emergency, either using battery-operated wall packs or smoke detectors with built-in light sources. These are not required in residential design but are especially desirable in an accessible home.

Controls are as important as fixtures in creating adequate light. Use pre-programmable dimmers to modify light levels. Consider using automatic sensors for hallways, turning lights on when a person enters. Choose keypad buttons with functions imprinted in Braille. Locate switches and outlets at consistent heights and locations so that a person with low vision or cognitive limitations will find them more easily. Use back-lit keypad or manual switches and dimmers to help find room controls in the dark. Finally, make sure that face-plate colors contrast with the wall background to improve visibility.

A multipurpose game room with a pool table also includes a piano, television, exercise equipment, sauna, drum set, and greenhouse. Over the long cold New England winters this space is much enjoyed by an active young couple and their community of family and friends.

> FOR MORE ON THIS HOUSE, SEE Case Study 13 (p. 160)

Sturdy wood and tile finishes and built-in cabinets with space for a wheelchair footrest show sensible long-term planning in this home for a young family. A raised hearth puts the fireplace within easy reach.

GAME ROOMS

For a family who bonds over cards or karaoke consider creating a game room. This should be an appealing place where people want to hang out, even without the games, as some will prefer to watch the gamers rather than play. Size the room around activities, which may include billiards, ping-pong, vintage pinball, air hockey, a table for board games, and multiple screens for video games. Provide ample storage for accessories and plenty of electrical outlets near electronic equipment to keep aisles clear of wires. Acoustic controls are critical here because the game room can get pretty noisy. Design environmental systems to maintain indoor air quality, as active groups raise air temperatures and humidity levels.

An upper-body workout machine in a great setting makes it easy for this homeowner to keep in shape. Combining the exercise equipment with other recreational activities is added incentive to use it.

The workspace and library share a hallway that links the kitchen and living areas, making it easy for parents to keep tabs on what is happening around the house.

HOME GYM

Home exercise spaces are more than entertainment in the accessible home; they are an essential component of the homeowner's physical therapy regimen. A convenient location makes it easier to exercise regularly, so locate the home gym near the bedroom or living spaces. Install a bathroom nearby, with a whirlpool tub for post-workout relaxation. Select exercise equipment that targets specific muscle groups. Consider how the homeowner's condition is likely to change, and plan for future equipment needs. Install solid blocking for wall-mounted or hanging equipment, such as a mechanical lift, and beef up the joists. Choose resilient flooring to reduce the chance of injury in a fall. Design the HVAC system for maximum control, as disabilities can reduce the body's capacity to self-regulate temperatures. A great workout space can make the difference between maintaining independence and losing muscle.

DESIGN COMFORTABLE SPACES FOR QUIET PURSUITS

Our homes shelter our quiet selves in addition to our sociable and active selves. Whether used for paying the bills or writing poetry, a study equipped with a desk and shelves is an integral part of many homes in this book. Quite spaces can be as simple as a sunny nook or yoga mat, or as complex as a library lined with built-in cabinetry. What's most important is that the home nurtures its inhabitants.

HOME OFFICES AND STUDIES

High-tech lifestyles have made the home office a place everyone wants to be. Today's workspace is used for managing household finances, for telecommuting, and for the kids' homework, and so it

With furnishings that are easily moved, this accessible home office above a garage converts to a yoga studio.

needs to be functional, adaptable, and comfortable. Consider the location carefully: either near family living areas or in a quiet part of the home, and on an accessible path of travel. In planning your own workspace, start by identifying equipment needs, such as standard computers, printers, and scanners, Braille panels, and new technologies such as touch-screen monitors, computers that respond to subtle cues such as a blink or a nod, and devices that convert between speech and writing. If the workspace is shared, be sure you have enough desk area and electrical outlets for dual equipment. For the well-organized home office, provide ample counters and sufficient storage for supplies (see the sidebar below left).

STUDY FOR A CHILD WITH LEARNING DISABILITIES

It's important to limit distraction and to provide an environment conducive to focused learning. Design a well-organized workspace with adequate desk area and storage nearby. Special-education teachers offer these guidelines for a homework center:

- Place the study desk against a blank wall and use a skylight or high window to bring in daylight.
- Provide ample storage nearby—hooks for backpacks and a jacket, shelves for books and notebooks.
- Subdivide storage drawers to reduce clutter and make it easy to find objects such as pencils or tape.
- Choose desk colors carefully, testing the colors with the user. Theoretically a white desk improves legibility, but people with dyslexia often find colored backgrounds distracting while reading. An object's color depends on context—its materials, the nearby environment, and the characteristics of the perceiving eye and brain—so what is legible to one person may not be to another. Control glare off the work surface by means of lighting and window coverings.
- Size and locate the desk for ergonomic comfort, matching desktop height and knee space to the user's own dimensions. (See the sidebar on p. 46.)
- Contain computer wires out of sight below the desk by leaving space for them, either through a hole in the desktop or a space between desk and wall.

"Students with learning disabilities can greatly benefit from a space specifically designed for homework and studying. When designing such a space, architects should help structure organizational systems, minimize distractions, and create clean and comfortable spaces conducive to learning."

—CASEY LAMB, SPECIAL EDUCATION COORDINATOR AT GIRLS PREP MIDDLE SCHOOL

YOGA AND MEDITATION AREAS

A calm space in the home for a regular practice can help prolong independence by maintaining flexibility. The yoga/ meditation room need not be large—6 ft. by 6 ft. accommodates a single yoga mat or a turning wheelchair—but it should have a serene atmosphere. If possible find a space with daylight (a window or skylight) or high ceilings for a sense of airiness, and away from noisy parts of the home, such as the kids' play space or a music area. Use acoustic construction to block unwanted sound. Use natural materials that are easy to keep clean—for example, wood, cork, or bamboo floors.

PROVIDE AREAS FOR CREATIVE SELF-EXPRESSION

Designing for a person's abilities is a powerful formula for creating an accessible home. Creative arts focus attention and develop coordination. Whatever artistic muses we honor, having a space to immerse in creative pursuits is an essential part of well-being for many people.

ART STUDIOS

Both artistic media and lifestyle influence the design of studio spaces, and certain needs are universal: durable finishes, plenty of storage, and strong, adjustable task lighting. Each craft also has unique requirements—for space, water, fireproofing, soundproofing, and utility hookups, as well as for media supplies, deliveries, and waste management. Design the studio around principles of ergonomics to reduce the physical stress of holding one position for a long time (see the sidebar on p. 46). Finally, find ways to showcase the artist's work, either as display or built into the studio or home.

MUSIC STUDIOS

For many people with and without disabilities, making music is one of the keys to living well. Whether used for jam sessions or recitals, the music room should be part of an accessible path of travel. Acoustical construction is especially important. You want to both keep sound from entering other parts of the home and prevent extraneous noise from intruding

AN ACCESSIBLE HOME work center meets both access codes and the user's own measurements. For a child in a wheelchair, tabletops should be 26 in. to 30 in. high and uppermost shelving from 20 in. to 36 in. above the floor. For adults, these figures are higher—tabletops at 28 in. to 34 in. and shelving from 14 in. to 48 in. Reaching over an obstruction such as a desk requires lower shelves. For the best fit, tailor designs to the actual user's dimensional requirements.

High windows and a northern orientation bring uniform daylight into this visitable art studio, created for a homeowner whose sister had polio as a child. With a large freestanding table in the center of the room and a long cabinet holding supplies, the room is always ready for the next project.

HOME ERGONOMICS

Proper body mechanics can make the difference between an injury and being pain-free at any age, as holding a difficult posture for a while is extremely fatiguing for the musculo-skeletal system. It's important to find a neutral position—comfortably upright, the head and vertebrae stacked vertically—that allows us to function comfortably. Each activity exerts unique strains on the body, whether sitting or standing, reaching, or bending. The parts of the body that are subject to stress from improper positioning are primarily the neck, shoulders, back, and wrists.

■ **Relax the shoulders and wrists.** Place work surfaces (keyboard, countertops, workbench) so the elbows are bent at a 90-degree angle, allowing the shoulders to relax. A too-high surface causes the shoulders to rise up around the neck and keeps the wrists flexed. Either position is exhausting.

■ **Relax the arms.** Avoid holding the arms at or above shoulder level for long periods of time. If you must reach upward to perform a task, such as a carpentry or art project, use a stepstool to bring your body to the area of the project.

■ **Relax the back.** Place the viewing surface (computer monitor, reading materials) just below eye level to be able to sit straight without leaning forward. Avoid surfaces (workbenches, sink bowls) that force you to bend forward at the waist. Sitting upright has the added benefit of opening up the field of vision, reducing eye strain.

To find the best position for activities in your life, take an inventory of the postures you hold for any period of time. If any of these are awkward, look for ways to alter your position by modifying the environment or using props. Moving often (stretching, changing position) is the best antidote to holding an awkward position. When we reduce muscle fatigue we reduce the possibility of injury and so make the home safer.

An accessible sound studio is used for music making and production.

A well-stocked woodworking shop close to ground level has wide barn doors for easy access and a large window wall to provide ample daylight for construction tasks.

on the music (see the sidebar on p. 39). Electrical requirements are equally critical for the room's success. Provide plenty of power for electronic instruments such as recording devices, speakers, and amplifiers, and keep wires out of the way by routing them into the floor or by installing raceways that keep them tight to the walls.

The idea of tailoring the home to the homeowners' lifestyles is not new, but it's an idea whose time has come, and the accessible home is charting new territory. If we can customize living spaces to suit each family's requirements, what can be done with the room where design choices often seem limited to granite vs. composite countertops? In the next chapter we'll look closely at the kitchen, and see how the accessible home is changing the way designers and homeowners look at this essential space. ✦

> FOR MORE ON THIS HOUSE, SEE
Case Study 19 (p. 189)

Accessible cabinet hardware can be used without a twisting or gripping motion. These custom-made door pulls in a home woodworking shop reflect the homeowner's personality and skill.

CHAPTER 5

preparing and cooking meals

LIKE ACCESSIBLE LIVING SPACES, the accessible kitchen is tailored to the homeowner's abilities and interests. Preparing meals can be a collaborative event or a chance to retreat into the zen of pepper-chopping. Whether you make your own pasta, shop regularly at artisan markets, or buy groceries in bulk, your kitchen needs are different from those of your neighbor who does take-out.

An injury or an illness initiates a crash course in managing with new abilities and limitations, a familiar process for people whose kitchens are featured in this book. Having

With a two-drawer dishwasher beside a two-bowl sink, cleanup can be done from a single seated position. Double-drawer appliances allow the user to reach in without having to move objects around.

"Design is directed toward human beings. To design is to solve human problems by identifying them and executing the best solution."

—IVAN CHERMAYEFF, DESIGNER

given a lot of thought to doing things with less effort, their experience is a tremendous resource for those planning a new kitchen. As a room where we spend much of our waking time, the kitchen has tremendous potential for making it easier to live with a disability. A successful kitchen is not about cabinet style or countertop materials—it's about the ways that each design can support its users impeccably. Let's look at how this takes place in the accessible kitchen.

PLAN THE KITCHEN TO BALANCE WORKSPACE AND STORAGE NEEDS

Kitchens demand meticulous planning because this is a space where inches count. Appliances come in fixed sizes (often fractions of inches), and cabinets are modular (often increments of 3 in.). A renovation project has existing walls and openings that limit usable space, and even an addition has constraints (such as lot setbacks and existing adjacent conditions) that limit what can be done. A typical small kitchen may have room for

This compact kitchen takes a minimum of space and works well with a nearby table.

Built-in appliances and pantry storage are consolidated on one wall to allow maximum openness at the counter workspaces in this kitchen for an elderly baker.

CHECKLIST OF USEFUL KITCHEN DIMENSIONS

Codes and regulations such as the Americans with Disabilities Act of 2010 set dimensional standards for accessible kitchens. Starting with these guidelines, check your own measurements for a good kitchen-user fit.

- Counter height 34 in., or an adjustable counter with a range of 28 in. to 36 in.
- Accessible counter workspace 30 in. wide.
- Clear approach space 30 in. by 48 in. in front of sink and cooktop.
- Knee clearance below sinks at least 27 in. high and 8 in. deep at the knees, 11 in. deep at the ankles (and lower for children).
- Toe clearances below sinks at least 9 in. high and 6 in. deep.
- At least 50 percent of storage within reach ranges, which are 15 in. to 48 in. high, except over countertops, where the maximum is 44 in. high.
- Appliance controls with a maximum operating force of 5 lb.
- Counter workspace 30 in. wide beside side-hinged oven doors and to one side of bottom-hinged doors.
- Be sure at least 50 percent of freezer space is within 54 in. of the floor. Provide a 30-in. by 48-in. clear floor space for approach, offset no more than 24 in. from the refrigerator centerline.

a double-bowl sink or a refrigerator with side-by-side doors—but not both. Each choice is a trade-off when space is limited. Approaching a kitchen remodeling project with a willingness to move doors, windows, and interior walls can mean the difference between gaining space where it is most needed and having a costly renovation with little to show for it.

Planning a kitchen means balancing requirements for storage (cabinets), workspace (counters), and appliances. The accessible kitchen adds one more requirement: putting everything within reach. One way to accomplish all this is by matching built-in items to the cook's own dimensions. The other is to create separate self-contained activity centers.

TAILOR KITCHENS TO THE USERS' DIMENSIONS

The first rule of kitchen planning has long been "design around the 36-in. countertop." Cabinets, dishwashers, and ranges are sized to fit under this counter so that the "average" user (a standing 5-ft. 10-in. male or 5-ft. 4-in. female) can comfortably cook. Many homeowners, however, actually prefer lower work surfaces, and now the ADA calls for 34-in. countertops. Setting the countertop heights early in the design phase will drive decisions as the project moves forward— from type of appliances to location of storage cabinets (see the sidebar at left).

It is a good idea to measure your own comfort ranges for working and reaching things in the kitchen. One homeowner purchased an adjustable table and spent several months trying out different small appliances at different countertop heights to find dimensions that worked for each member of the family (see Case Study 4, Living Big in a Small House, p. 115). Take an inventory of your own cooking habits. Do your meals tend toward complex dishes presented with artistic flair? If so, you'll want extra work area and space to store serving platters. Do you prefer one-pot meals such as hearty soups and stews? Plan a well-stocked cooking center with extra pull-out shelves for chopping. Do you bake bread or decorate cakes? You'll want lower countertops. Knowing your cooking preferences will help you design a workspace that meets your needs. The principles of ergonomics (see the sidebar on p. 46) make it possible to establish installation heights that reduce body strain and improve comfort while cooking and preparing meals.

DESIGN THE KITCHEN AROUND ACTIVITY CENTERS

The second rule of kitchen planning has been the supremacy of the "work triangle"—a layout with the sink, fridge, and stove all a short walking distance apart. For the accessible kitchen, however, a more useful model has several work areas, each designed so that everything needed can be accessed from a single position. These activity centers can be different for each homeowner. Consider providing a morning coffee center, a salad counter, a baking area, a children's snack area, and a pass-through between the garage and pantry for transferring groceries. A well-planned kitchen can last decades—that's over 1,000 meals each year!—so it's important to take the time to get the layout right.

> FOR MORE ON THIS HOUSE, SEE Case Study 6 (p. 125)

With a single pole for support, the island countertop has knee space below for seated cooking, mobility devices, or casual eating.

BUYING AN
APPLIANCE
If you can't visit a
showroom to test
out an appliance
before purchasing
it, then find a
supplier who will
allow you to return
an item that isn't
a good fit.

Raised knobs with
indicator bars make
cooktop dials easy
to operate with weak
hand control. An
appliance with knobs
at the front edge
ensures that the
cook never has to
reach across flames
to turn off a burner.
For people with low
vision the contrast
between burners
and cooktop makes
it easier to see the
work surface.

SELECT APPLIANCES FOR SAFETY AND EASE OF USE

Appliance manufacturers have gotten the message that today's homeowners want choices, and options now include accessibility features. When shopping for appliances choose those that place food preparation areas within reach and that are easy to operate. Consider appliance door swings when laying out the kitchen, and make sure the configuration let's you look inside without contorting the body. Here are some features to look for:

- Shallow dishwasher and freezer drawers reduce the amount of bending to reach inside.
- Shallow sinks let a seated user reach the bottom of the bowl.
- Drain holes and plumbing traps at the back of the sink bowl leave knee clearance for seated users. Insulate or enclose the plumbing to protect legs from hot pipes.
- Single-lever faucet controls, located at the side of the sink, offer low-effort operation.
- Instant-hot-water dispensers shorten the steps involved in making a hot drink.
- Cooktops separate from wall ovens allow both to be placed at comfortable heights.

- Magnetic-induction cooktops keep the surface cool while food is cooking.
- Cooktop controls at the front edge prevent reaching across burners to adjust the heat.
- Ovens with side-swinging doors let a seated cook get close to the opening.
- Automatic shutoff controls and gas alarms are safety features if you leave the heat on.
- Use a 34-in. range if countertops are 34 in. high.

These features make it easier for people with a variety of impairments to work in the kitchen, but they are not enough to make sound choices. You need to look a little closer to see if an appliance works well for you. Even with products labeled "ADA-compliant," homeowners need to perform due diligence, and principles of universal design (see the sidebar on p. 12) offer a good framework for evaluating appliances. Consider comfort in addition to capacity. Is it easy to use? Are the operating instructions clear? Is it easy to clean? Online shopping has its limitations, as the information the buyer wants is not always offered.

As important as style and performance may be in selecting appliances, the design of controls has a greater impact on everyday use. Whether you rely more on sight or touch to operate an appliance, whether you sit or stand while working in the kitchen, some controls will be easier than others. A simple on-off device is an anachronism in the world of smart appliances, programmed to have dinner ready when you return home from work. Select tactile controls, such as raised buttons or dials with directional indicators that click into position at each setting. Touch-pad operation is great for busy hands and arthritic fingers but tough on weak eyes. Adding bubbled surfaces or a clicking action makes the touch-pad

> FOR MORE ON THIS HOUSE, SEE Case Study 24 (p. 216)

This appliance garage shares a room divider and serving counter. With its floor at the same level as a built-in table, appliances can easily be pulled into position for use.

more useful (see the sidebar at right). If the controls can be operated manually but programming activities are especially complex, another set of problems arises. As you select appliances for your home, consider user interface, from programming through a wide range of cooking activities.

The same questions of legible information and usable controls apply to purchasing small appliances as to large. Microwave, toaster, toaster oven, slow cooker, coffeemaker, espresso machine, bean grinder, mixer, blender, food processor, bread maker, juicer, soda

Built-in cabinets for microwaves and wall ovens should be placed within the cook's comfort range for safe operation. White push-buttons on a black background make an appliance accessible for those with low vision.

KITCHENS FOR VISION IMPAIRMENT

Individual experience, training in blindness skills, tactile acuity, and temperament all play a role in the choice of products. Choose white appliances and contrasting black knobs so that controls are easy to see. Avoid controls with silvery or reflective surfaces, which blur printed information. Use a dishwasher with contrasting-color interior racks, or ovens with light-color interiors and easy-clean settings. Select a refrigerator with uniform interior lighting. Consider appliances with voice-recognition or talking controls.

Design countertops for safety and comfort

Provide ample countertops so that items can be left in place for easy retrieval. Round countertop edges and corners to reduce the chance of injury as someone bends down to feel for objects dropped or stored low. Choose countertop colors to contrast with appliances and utensils to make it easier to distinguish foods and features. Keep the cooktop away from windows to reduce glare that can obscure views of the work surface.

Compartmentalize storage cabinets

Food identification relies on container sizes and shapes (canned soups vs. tuna vs. pet food, large/small, square/round), so coordinate storage compartments with your grocery list. Install drawer dividers and add extra shelves to avoid high stacks of stored items. Make shelves shallow—one box or can deep. Subdivide drawers to make it easier to find objects by touch. Place adequate storage within reach to avoid using ladders to access high cabinets. Place ample storage near appliances—for example, pot holders near the oven, spice racks near the cooktop, coffee supplies near the coffeemaker, baking supplies near the mixer. Adequate storage eliminates clutter by making a place for everything.

SMART APPLIANCES

For people without full use of their hands, eyes, or ears, smart appliances promise to simplify everyday tasks in ways unimaginable until recently. Many companies are producing devices that monitor energy consumption or that convey diagnostic information to a technician so that repairs can be made accurately and quickly. Embedded microchips will bring an explosion in gadget capabilities, making it possible for home appliances to connect with the web and also with one other. As you consider making major purchases for your home, consider ease of use as well as function.

Putting away groceries is easy when food storage is concentrated in a generous pantry.

maker, electric knife sharpener, and can opener—all these objects require space in the kitchen. It's not unusual to see small electronic devices scattered about the counters in a new kitchen, seemingly afterthoughts in the design process. When small appliances are identified at the outset, they can be located where they are most useful and stored where they can be reached. The accessible kitchen makes life easier by ensuring that everything has a place.

DESIGN CABINETS FOR EFFICIENCY AND FUNCTIONALITY

It's easy to be wowed by beautiful wood finishes when shopping for cabinets, but first, a little reality check: Cabinets are basically boxes with doors, so it's what's inside that counts. Standard millwork consists of base cabinets with a shallow drawer above roomy shelves, and wall cabinets with adjustable shelves. Most manufacturers offer accessories to make the accessible kitchen more convenient, so we have cabinet fit-up options such as all-drawer bases, drawer dividers, tray slots, wine and spice racks, and corner

Side-accessed pull-out drawers pack a lot of food into this food pantry. The front panel of each cabinet mirrors adjacent drawers to give a sense of proportion and scale to the kitchen.

lazy Susans. There is great variation in pantry and trash features, so consider your shopping schedule and your town's recycling program as you plan the kitchen. By choosing semi-custom or custom cabinets, you are buying increased ability to tailor the cabinets to your space and requirements.

The room configuration dictates what can be done with a kitchen. Effective kitchens can be designed using the U-shape, the L-shape, or the galley layout, as long as there is adequate maneuvering area between cabinets. Island and peninsula schemes both provide a central workspace used for face-to-face cooking and serving while acting as a buffer to adjacent living areas, and most homes seem better suited to one or the other layout.

Accessible cabinets generally have looped pulls rather than knobs, operable with a closed fist and without twisting, grasping, or pinching motions. Drawer glides and door hinges are standard features yet there are still choices to be made regarding how these work in an accessible kitchen. An increasing array of special options promises to make the kitchen even more accessible (see the sidebar on p. 56).

Two parallel banks of cabinets with an accessible aisle make a galley kitchen with plenty of maneuvering space.

Low drawers make it easy to reach items stored inside from a seated position. Sturdy drawer glides and deep drawers accommodate heavier storage items. Pull-out shelves under the counter add extra workspace where it's most needed.

THE BEST PLAN for your kitchen is the one that provides the right balance between countertop area, maneuvering space, and storage square footage. By comparing various layouts with your present kitchen's capacity, you can find a plan that meets your needs.

HARDWARE BELLS AND WHISTLES

Visiting a kitchen showroom, it is easy to be convinced that the main choices involve cabinet face details, finishes, and pulls. While drawer glides and door hinges are standard fare for cabinet hardware, installers can often make substitutions that improve accessibility. For example, to keep the path of travel clear, choose hardware that lets doors and drawers shut completely. To reduce extraneous noise, chose silent closings and door bumpers. To make the best use of reach ranges, specify pull-down shelves and pull-out counter extensions. Some of the hardware options are as follows:

- **Full extension drawers** pull to the face of the cabinet, shortening the reach inside.

- **Soft-close fittings** shut doors completely to keep the path of travel clear.

- **Silent-closing fittings** close drawers without slamming to protect stored items.

- **Door shock absorbers and bumpers** quiet the closing sound and protect surfaces.

- **Touch-release (or push-latch)** drawers and doors use light pressure to open and shut, avoiding surface-mounted pulls that can get in the way.

- **Offset and pivot hinges** let doors clear the cabinet opening when fully extended.

- **Push-button controls** make it easy to open cabinet doors (but they need electricity).

- **Upward-acting cabinet doors** slide up out of the way (high ceilings are needed).

- **Swing-up hardware** allows wall cabinet doors to hinge at the top and stay open.

- **Pull-down shelves** pivot forward and down to bring upper shelves within reach.

- **Catches** (magnetic, ball, roller, or spring-operated) keep cabinet doors from drifting open.

- **Pull-out tables** fold back into the cabinets when not in use.

- **Foldaway systems** support sliding shelves for small appliances.

Be sure to have your cabinet supplier coordinate the fabrication and confirm that hardware is suitable for the intended use. Door weight and hardware can impact cabinet size, so some decisions may need to be revisited before finalizing purchase and installation. As a homeowner, you need to give the cabinet fabricator your performance requirements—such as what you will be putting in a drawer, weight of items, available space dimensions—and let him or her choose the appropriate hardware.

> FOR MORE ON THIS HOUSE, SEE
Case Study 4 (p. 115)

The smooth, clean lines of flush cabinets seem to belie the complex planning that went into this kitchen. An appliance garage with a retractable door puts the coffeemaker and toaster in easy reach—and out of sight when breakfast is over. A sliding drawer in hard-to-reach corner cabinetry was adapted for side access.

DESIGN COUNTERTOPS FOR COMFORT AND PRACTICALITY

To be fully functional, workspace needs to be large enough but also well placed (such as to each side of a sink and cooking surface, and near a wall oven and refrigerator). A counter off in a far corner of the room or beside the back door is likely to become a home office or catch-all rather than a workspace for preparing meals. *What* is provided is only part of the story; *where* and *how* are equally important in the accessible kitchen.

Designing countertops means selecting materials and planning the details. Use heat-resistant materials around cooking appliances and avoid mildew-prone materials in wet areas. Solid-surface countertops such as stone, Corian®, and composites can be contoured into a grooved drainboard beside the sink.

Metals such as stainless steel and copper can be formed with a raised perimeter edge to keep foods from spilling onto the floor and creating a slipping hazard. Supplement the fixed countertop with pull-out shelves in areas of heavy use, such as the cooking center or the prep sink.

CONTROL SAFETY AND COMFORT WITH AN ELECTRICAL PLAN

Good task lighting improves kitchen safety by making it easier to see what you are doing (see the sidebar on p. 41). Each kitchen has many identities—a bustling prep center, a messy clean-up center, an invigorating wake-up center, a place for late-night heart-to-heart talks. One goal of lighting design is to vary the light levels to suit these identities. For a gentle glow on the floor as you shuffle in for a late-

COUNTERTOP HEIGHT By setting countertops at different heights the work areas can serve different cooking functions—say, a lowered pastry counter or breakfast bar—and different chef sizes.

This kitchen has four countertops at four different heights. Full-height cabinetry at the end wall holds additional workspace—small appliances pull into position on sliding shelves, hidden behind doors that slide out of the way when the appliance is in use.

Drawers and a continuous towel bar make this industrial kitchen counter especially functional. The stainless-steel countertop makes a durable surface for transferring heavy pots of boiling pasta water between the cooktop and sink (to the left). A raised edge keeps spills off the floor.

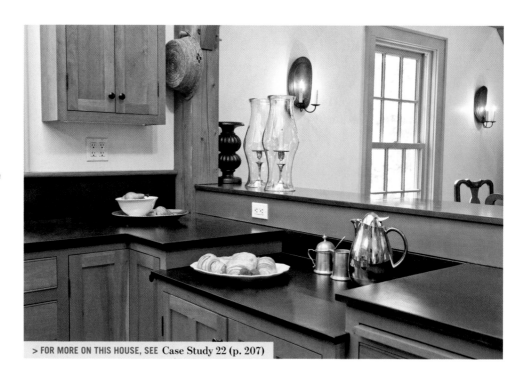

> FOR MORE ON THIS HOUSE, SEE Case Study 22 (p. 207)

The low wall separating kitchen from dining area makes a good staging area for serving and keeps the kitchen clutter out of view from the table beyond. Countertops at varying heights serve different purposes, from baking to serving.

OUTLETS AND FACEPLATES
Incorporate outlets on cabinet faces to shorten the reach for plugging in small appliances. Contrast electrical faceplate colors with wall colors to make them easier to see.

night snack, go with ambient lighting. For bright lights on the cutting board surface, go with task lighting. Dim the lights to divert attention from dishes piled high in the sink, and use accent lighting to focus attention on your refrigerator art gallery. Another goal is to locate the lighting controls so you never have to navigate in the dark. Multi-pole switches allow you to turn on kitchen lights whether you arrive from the bedroom at night or the yard at dusk.

A lighting plan considers both how the light falls on surfaces and also how the fixtures appear alongside other installations. This is accomplished by selecting different light fixtures for different purposes. Choices include recessed lights, under-cabinet fixtures, bulb-strips, cove lights, wall sconces, pendants, spotlights, step lights, and ceiling fixtures. Location is as important as fixture selection to create a well-lit kitchen that promotes cooking safety. By centering fixtures on windows or over countertops and aligning them over aisles,

the ceiling complements the floor layout and puts light right where it's most useful.

Consider how many electric outlets you'll need—and then add more. Someday cordless appliances may be the norm, but for now, kitchens need enough power for all our devices. Installing extra electrical outlets makes it easier to add small appliances when gadget designers come up with the next foodie toy. The kitchen should have enough outlets, located in areas of use, to avoid having electric cords trailing in the way.

DESIGN LOW-MAINTENANCE KITCHENS

Maintenance tasks are particularly onerous when the homeowner has a disability, and so the accessible kitchen should be easy to keep clean as well as in working order. Choose cabinets that are sturdy and well made. Select adjustable hardware to keep cabinet doors in place, and pulls that are comfortable to the touch. Choose dimensionally stable countertop materials that are impervious to mold and mildew, scratches, burns, melting, and cracking. Use flooring that is level and nonslip, with stain-resistant, durable finishes. Specify washable wall surfaces and fade-resistant paints. Choose light fixtures with long-lasting bulbs, positioned for easy re-lamping. A low-maintenance kitchen is a low-stress kitchen, qualities compatible with the accessible kitchen.

DESIGN FOR FUTURE ADAPTATIONS

By constructing flexibility into the plans, modifications can be made at a later date with a minimum of disruption, time, and cost. One way to accomplish this is with 30-in. base cabinets that can be removed to make space for seated cooking in a few areas of the kitchen; be sure to continue floor and wall finishes behind the cabinets to avoid the cost of patching-in later. Rolling base cabinets have built-in

flexibility—with finished tops they can be used for extra workspace or as serving trolleys and tucked back under the main counter when not in use. Rough plumbing installed under cabinets makes it easier to add sinks in the future.

Whereas the kitchen is a place where we care for others, the bathroom is a place where we care for ourselves. Both are places where accessibility adaptations are demanding and essential. Access codes offer guidelines for making these rooms functional for people with disabilities, but little inspiration on making them wonderful places to be. Let us turn now to the accessible bathroom to see how it can make life with a disability easier. ✦

A base cabinet on wheels doubles as a rolling trolley for transporting hot dishes from countertop to table, or for breakfast-in-bed. Legs support the countertop when the trolley is in use.

CHAPTER 6

personal hygiene and care

BATHROOMS ARE DANGEROUS places, and we all put ourselves in jeopardy several times a day. More than two-thirds of emergency room visits are due to bathroom falls. For young adults, the tub and shower areas are most hazardous. For the elderly, most falls occur near the toilet. As we look at creating accessible bathrooms we need to find ways to make this space safer for everyone.

What causes all this falling? We slip on scatter rugs and wet floors. We lose our balance while bending down to pick up a dropped bar of soap. Heat from the shower or tub water causes

A wall-hung sink gains storage with a stepped-wall behind it.

"Design is not just what it looks like and feels like. Design is how it works."

—STEVE JOBS

> FOR MORE ON THIS HOUSE, SEE **Case Study 6 (p. 125)**

Two children's bedrooms share a common bathroom in this home where the mother has a physical disability. With an accessible bathroom, their mother can help them with personal care tasks, or use this bathroom if her own is occupied. A shallow sink, low mirror and windows, and generous toilet transfer space all improve accessibility.

dilation of the blood vessels, which, in turn, can lower blood pressure and cause a person to become light-headed and therefore vulnerable to a fall. Feeling a little dizzy, we grab for anything nearby, usually a towel bar or a slippery porcelain sink, neither of which is designed to carry human weight.

If there is a grab-bar in the shower, we can't get a good grip with wet soapy hands. When we lean back to sit on a low toilet, especially if the leg muscles have any weakness, and let our bodies drop those last few inches—boom!—we miss the seat. When steam in the bathroom fogs the mirror above the sink, we lose the sense of balance that comes from having a clear sense of where we are. Falls are more hazardous when cast-iron heaters with rigid edges up the ante on hitting the wall. Since activities of daily living can be challenging for people with disabilities, and since the bathroom is the site of so many Activities of Daily Living (ADLs), it is critical that the bathroom designer make the effort to create an exceptional space.

DESIGN THE BATHROOM FOR ALL ITS USERS

Each person has a unique way of managing personal hygiene, and the designer who knows what is involved can accommodate the necessary activities. Intimate discussion is therefore a must. Here are some questions to get the ball rolling when someone in the family has a disability:

- What bathroom activities, if any, require assistance? You may need to size the room for several people.
- What are the homeowner's medical needs, and what apparatus may be required? Catheterization supplies, for example, may need to be stored near the toilet.
- How predictable are bodily elimination functions? Install a bidet or shower near the toilet, or make sure there is space for cleanup and sanitary supplies. A larger shower drain may be appropriate.
- Are tubs or showers preferred? You may have space for one or the other, but not both.
- Which muscle groups are reliable? Strong biceps favor grab-bars located for a pull-up motion, and strong triceps favor a push-up motion such as leaning on a low wall.
- What are the user's manual skills? Consider no-touch controls for the sink and toilet. Left- or right-handedness guides placement of accessories and transfer logistics.

- What side of the toilet is farthest from the wall? Make sure the flush valve is on that side.
- Can the body self-regulate temperatures? Activities take longer with a disability, and some conditions (such as spinal-cord injury) disrupt the body's ability to adapt to changes in heat and cooling. Environmental controls are especially important.
- What is the prognosis? If the condition will deteriorate over time, you will need to make space for both the user and a caregiver.

Designing a bathroom that does not feel institutional is a goal shared by many. Create adequate storage for medical supplies so these do not overtake the counters. Integrate grab-bars and lifts into the lines and character of the bathroom. Look for ways to repurpose products from other fields and to re-configure items from health-care settings (see Case Study 14, Bathroom for Mother and Son, p. 165). Choose attractive and accessible residential fixtures. Install complementary sinks for standing and seated users. Mount mirrors that everyone can use—either tilting or flat and full-height. Engage the family in creating a bathroom that meets everyone's needs.

Are there sensory impairments that need to be considered? A person with visual limitations needs extra light, subdivided storage compartments that give each object its proper home, and contrasting colors that make it easier to distinguish sink from countertops. In a home where someone is hard of hearing, locate the light switch on the hall side of the door, so that those waiting to use the bathroom can announce their presence when a knock isn't enough. For someone with cognitive impairments, limit distraction by providing enough storage and workspace so that counters are not

Low shelves rather than grab-bars were designed so that a person with low hand-strength can lean on forearms rather than hands to support body weight during transfer to and from the toilet.

cluttered. And if reflections cause anxiety, remove mirrors or enclose them (for example, behind a custom-design cabinet panel).

PLAN AROUND THE USER'S MOBILITY NEEDS

The accessible bathroom has a 30-in. by 48-in. space for mobility devices in front of each plumbing fixture and room to turn around in a wheelchair. It has handrails for security, unless sturdy cabinets or fixtures can be located in a way that allows a person to "furniture-walk" (hold onto objects in the environment, much like a toddler "cruises"). Items needed for each functional area are stored near places of use (see the sidebar at right). The homeowner may want to keep some items out of sight, especially if the bathroom is used by guests or other members of the family.

USE WITH TRANSFER EQUIPMENT

Knowing how transfers take place between assistive devices and plumbing fixtures lets the designer plan accordingly. You may need to install equipment such as a mobility lift within the bathroom or with a link to the bedroom. Convenient wheelchair parking space should be provided either at one end of an overhead lift or where the transfer occurs.

ASSISTED USE

Bathrooms need to be a little larger when a person has a mobility device (wheelchair) plus an assistive device (overhead lift) and a caregiver. Be sure there is enough space for two people and equipment at each fixture. To capture a few inches here and there, customize built-in items such as the vanity counter or shelving beside the toilet. Double up on hand-shower fittings so both the user and caregiver can have a hand in washing. Good lighting is

A PLACE FOR EVERYTHING

Yesterday's catch-all storage compartments—the hall linen closet and in-wall medicine cabinet—do not serve today's lifestyles. The well-planned bathroom has storage designed for convenience and aesthetic compatibility. The accessible bathroom raises the bar; each functional area has workspace and storage tailored to the user's needs and within reach, without moving from areas of use. Shelves and cabinets, hangers and drawers—design of storage compartments should be fine-tuned for the intended purpose, subdivided and well-lit. Consider the list below when designing bathroom storage:

- **Shower area:** Hair-care products, shaving supplies, rolling shower seats.
- **Tub area:** Soaps and face-cloths, hair-care products, rubber duckies, reading materials.
- **Toilet area:** Extra toilet paper, sanitary products, wipes, medical equipment, portable lift motor and sling, transfer board, reading materials. (Consider homeowner interests: I once had a client who played the guitar while on the toilet!)
- **Sink area:** Dental-care products, prescription and over-the-counter medicines, eye-care and contact lenses, ear/nose/throat products, first-aid supplies.
- **Vanity counter:** Hair dryers/curlers/flatteners, hair-care products, accessories, clips, and make-up.
- **Other storage:** Bathroom linens, room-cleaning supplies.

Simple open shelves put storage just where it is most needed for the sink and toilet.

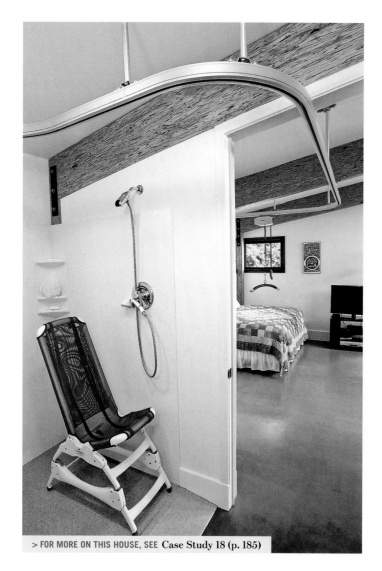

> FOR MORE ON THIS HOUSE, SEE **Case Study 18 (p. 185)**

A mobility lift utilizes an overhead track to connect the bed with the bathroom. Sealed concrete floors and a synthetic shower surround with a tapered floor and drain combine for an easy-maintenance bathroom.

important, and should be planned to avoid shadows, particularly in areas of clean-up.

DESIGNATE A VISITABLE POWDER ROOM FOR GUESTS

Along with an accessible entrance and path of travel, the visitable home must have at least one powder room on the main floor (see the sidebar on p. 16), usable by a person in a wheelchair. The 1996 Fair Housing Act offers useful guidance as to what constitutes a visitable powder room. It must be large enough for a person to enter, shut the door, use the fixtures, re-open the door, and leave the room. Room size is a function of window size and placement, door location and swing, plumbing fixture dimensions, and direction of travel. Provide a 30-in. by 48-in. space for a wheelchair at each fixture, clear of interference from a swinging door. The toilet needs a grab-bar to one side, and a distance of 18 in. to the nearest wall or fixture on the other side. The sink can be wall-mounted or set into a vanity counter. Approached from one side, the vanity sink can have cabinets below, as long as the required wheelchair clearance is met within the room. Approached from the front, the sink needs a 34-in. maximum rim height and at least 27 in. for knee clearance, and removable cabinets below the sink.

A powder room feels spacious with a long vanity counter and mirror. Base cabinets provide ample storage for linens and supplies. A shelf beside the toilet, shown in reflection, ensures that each functional area has what it needs.

Like other accessible rooms, the visitable toilet room needs at least a 32-in. door clearance, and the door may swing either in or out. Provide storage for extra paper rolls near the toilet. If the first floor has a full bathroom then either the tub or shower, but not both, needs to be accessible. If there are two sinks, one needs to be accessible.

This curbless shower in the family bathroom increases space for bathing as well as for the adjacent sink and toileting areas.

DESIGN CURBLESS SHOWERS FOR SAFETY

Curbless showers can be used by everyone. By shaving off the top surfaces of joist framing below, shower floors can be sloped to drain. Collapsible rubber dams can be installed to contain water in the shower. Size the shower for either a transfer seat (36 in. wide) or for wheelchair roll-in (60 in. wide, large enough for turns and assists). Design shower seat heights for the user's comfort, usually 17 in. to 19 in. high.

Shower accidents are five times more likely to occur while getting out rather than getting in, and are disproportionately common among people ages 15 to 24, so sturdy grab-bars should really be in all shower areas (see the sidebar on p. 60). Other thoughtful details help make the shower safer. Choose nonslip floor materials such as textured tile or a slatted wood tray over a concrete floor. Provide adequate storage so that hair-care and shaving products are less likely to fall on the floor. Install glass walls and overhead lights to improve visibility within the shower. Locate a towel shelf or hook within easy reach for drying off before exiting the shower. Place shower controls so they can be operated without the user getting wet, and locate a rolling shower curtain above a sloped floor so that water is deflected back to the drain.

Unpredictability around bodily elimination suggests having easy-to-use clean-up areas near the toilet, and showers

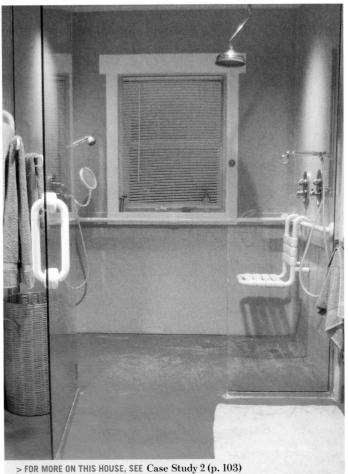

> FOR MORE ON THIS HOUSE, SEE Case Study 2 (p. 103)

A tempered glass shower enclosure with a double-swing door makes it easy to see if someone has fallen while showering. Grab-bars on all three shower walls are aligned with the vanity to the right for a clean horizontal aesthetic.

GRAB-BARS

A person falling will grab hold of anything nearby, but towel bars and robe hooks are not engineered to support human weight. This is starting to change, as new accessories are being designed for dual purpose, such as corner shelves with integrated handrails. Attractive grab-bars that match room finishes are an aesthetic we will be seeing more and more in our homes as builders incorporate universal design features. It's not enough that the grab-bar be sturdy: The installation needs to be securely anchored into the wall framing. Whether or not you choose to install grab-bars in the bathroom now, you can make it easier to install them later. Plywood sheathing 4 ft. high on all bathroom walls gives flexibility for locating bars. For a growing child or a person with a progressive disability, planning now will save money and time later.

Specify grab-bars that are 1½ in. to 2 in. diameter, either round, oval, or square with eased edges and with a knurled, nonslip surface. Mount the bars 1½ in. from the wall to prevent entrapment. Install bars to withstand a 250-lb. load in any direction. Grab-bars beside the toilet may be fixed or folding, and wall- or floor-mounted. Set bar heights within reach, and within ranges set by access codes, which call for grab-bars in several locations:

- **Toilet:** Side and rear walls.

- **Tub:** Two bars on the side wall at standing and sitting reach ranges, one bar at the foot (control end) and one bar at the head if a removable seat is used.

- **Shower:** All three walls in a roll-in shower, and two walls in a transfer shower (beside and in front of a seat)

TRAP PRIMER Keeping indoor air quality fresh in a room with a floor drain can be a challenge if the drain is allowed to dry out between uses. Install a trap primer in the plumbing below the floor. It will prevent sewer odors from backing up into the house by keeping the trap filled with water.

The master bathroom has a spacious tub and shower area, with plumbing controls set within reach at each fixture. Use of a linear rather than circular drain allows the floor to be evenly sloped to reduce the chance of tipping the wheelchair during turns.

fit the bill, especially when the floor drain is large enough to collect solids. Sloping the whole bathroom floor slightly toward an in-floor drain makes room clean-up easier. (It even helps with washing the dog!)

CHOOSE STEP-IN OR WALK-IN BATHTUBS FOR COMFORT

There are so many choices for tubs that it should be possible to find the right balance of therapeutic features, comfort, and accessibility, but due diligence is advised. As the bathroom is constructed around the tub, this choice is not easily undone. Standard tubs are shallow and often need adaptations to be accessible for people in wheelchairs. Hydraulic chairs installed inside the tub pivot for transfer and then lower the body into the water. Sling-type lifts on overhead tracks perform a similar function, carrying the bather in from the toilet or bedroom. Rolling transfer benches adapt the tub for showering. Match bench functions with the user's capabilities: Avoid benches designed for the center of the tub if the user is unable to bend at the knees.

Another option is the drop-in tub, set into a platform that serves as a transfer bench and allows plumbing controls to be placed at the front edge. Soaking tubs have a smaller footprint but are 6 in. to 18 in. higher than standard tubs, and so may require a step. Consider hydrotherapy in choosing a new tub; a whirlpool tub sends pressurized water for deep-tissue massage, and an air-jet model sends warm air bubbles for a soothing experience.

New products address a variety of bathing needs. Wheelchair-accessible tubs include a model designed for transfer to a seat, with a side wall that slides up a vertical track for entrance (see the photo at right). Tubs with a raised base accommodate the wheels of a rolling lift. For people who are able to stand and take a few steps, walk-in tubs are a good alternative. Most walk-in tubs have a narrow hinged door and a heated seat, which keeps the bather warm as water is filling or leaving the tub. Some companies retrofit acrylic tubs for use as seated roll-in showers.

It is advisable to visit a plumbing supply showroom to test drive products before purchasing. By trying out various fixtures and fittings the user can find products that are both comfortable and convenient. Choose a faucet, a drain with trip-lever and overflow, and lever controls that are easy to use with a single hand motion and closed fist. Choose a showerhead with adjustable controls to vary the flow of water and also an adjustable-height hand shower for seated bathing. Install tub controls near the front edge to avoid a long reach. Coordinate the controls and hand shower with grab-bars, and mount everything at convenient heights on the tub wall.

Simple modifications made the original bathroom in an old house accessible. A clawfoot tub was replaced with a soaking tub, and sink storage was expanded with drawers and a wrap-around countertop. Openings in the drawer faces are easier to use with weak hand strength than pull-type hardware.

PLUMBING TURN-ONS

- Use anti-scald, pressure-balanced mixing valves to maintain safe and comfortable water volume and temperature during sudden changes in water pressure.
- Limit water temperatures in the tub and shower to 120°F (49°C).
- Provide a hand shower with a hose at least 59 in. long, making sure the spray is capable of delivering the same water pressure as a fixed showerhead.
- Select large lever-type controls that are easy to see and operate.
- Choose multifunction showerheads that spray water in a fine aerated mist or a vigorous revitalizing massage.
- Consider mounting body-sprays along shower walls for all-over washing.

This visitable bathroom has an accessible sink and a spacious transfer area beside the toilet. The toilet was specified with a flush-lever on the right side, easily used from the open part of the room.

Deep cabinets are sized for both toilet and sink supplies, and keep space near the toilet open for transfer. Open vanity counters for two mean double the space for oral and hair-care products.

> FOR MORE ON THIS HOUSE, SEE
Case Study 6 (p. 125)

DESIGN TOILETS FOR INDEPENDENCE

A toilet that is easy to use fosters independence. Seat height and bowl configuration are two important comfort factors, and most standard seats (14 in. to 15 in.) are a little low for people to get on and off safely. "Comfort height" toilets raise the seat by a couple of inches, and wall-hung models can be mounted at a height that works for the user while keeping the floor below open for easy cleaning. Toilets come with elongated and round bowls, and seats that slope inward to control spills. One person's comfort is another's cramped legs, so if you can, find a showroom where you can try sitting on various models.

The bathroom layout should provide adequate space for approach and transfer—whether one transfers from the front or side, diagonally or parallel, or uses a rolling seat or lift. If space allows, consider installing a Washlet®, a toilet seat designed to spray water and a flow of warm air for sanitary self-cleaning. Specify the flush-valve location carefully, as most toilets will have these on the left side as you face the toilet. You want to avoid having to twist and reach around the toilet, which is another falling hazard.

The toilet area should have everything needed nearby. Mount continuous-flow toilet paper dispensers at a comfortable height and forward of the bowl. Provide grab-bars, either fixed or fold-down, or low shelves to guide the user into position and assist in getting up. The designer who knows the homeowner's toileting logistics can ensure that storage is properly sized and located.

DESIGN SINKS FOR CONVENIENCE

Accessible sinks have clear space below for knees and footrests for a person in a wheelchair, with a 30-in. by 48-in. approach space to the front. They also have surfaces nearby for all the paraphernalia that is used near the sink. Vanity counters with drop-in sinks fill the bill in the accessible bathroom. Locate the vanity at a comfortable height above the floor— 29 in. to 34 in. for seated use—and use a shallow sink for easy reach into the bowl. Locate lever controls beside rather than behind the sink. Enclose or insulate plumbing pipes to protect legs from being injured by hot pipes. For an adaptable bathroom, design removable cabinets below the countertop, with floor and wall finishes continuous under the cabinets.

Whether a vanity or a wall-mounted sink is used, place storage to one side or nearby, and add generous counter work-space. Wall mounted sinks should have hangers for secure mounting, to avoid failure when people lean on the sink for support—another avoidable hazard. Swing-arm sinks offer the ability to adjust the height or location for convenient use. Use levers or no-touch fittings for people unable to use their hands. Provide

Open knee space below a cantilevered sink combine with tall pull-out drawers, reachable from both sides, to make an accessible and functional bathroom vanity.

Conveniently located between sleeping and living areas, this bathroom can be accessed with minimal effort. Wide pocket doors allow the bathroom to become part of either space.

DOUBLE SINKS
When two people
share a vanity
counter, consider
going with
complementary
sinks—one drop-in,
the other raised—
sized for seated
and standing use.
To avoid making
people bend
forward, measure
comfortable reach
ranges for each user
before setting sink
heights.

ample electrical outlets for personal-care appliances, located within easy reach, such as on the apron in front of the vanity counter.

PROVIDE NECESSARY ACCESSORIES

Mount or tilt mirrors low enough for viewing from a wheelchair. Mirrors over sinks should have the lower edge not more than 40 in. high, and other mirrors should not be more than 35 in. above the floor. A full-height mirror should be placed somewhere in the home, and the bathroom is one good location. Install mirror defoggers—electrical fabric that keeps mirrors warm—for clear mirrors when the room gets steamy. Select robe hooks, towel bars, paper dispensers, soap dishes, toothbrush holders, and shower shelves in sizes and styles compatible with the bathroom design and mount these within easy reach.

Bathrooms can be bright and reflective spaces, with mirrors, chrome plumbing fittings, and glazed tile, so getting the lighting right without causing glare can be a design challenge. Dimmers give you the best of all worlds: brightness with softness. As in other areas of the home, task and ambient lighting each serve a different function and should both be provided (see the sidebar, Lighting the Home, on p. 41).

Use a quiet exhaust fan sized to fit the room cubic footage, and separate it from the lighting controls to reduce ambient noise in the bathroom. A timer for the exhaust fan conserves energy by making sure the fan turns off after a given time—a nice feature for people who may forget to flip the switch. For a home where people have hearing loss, place lighting controls on the hall side of the bathroom door, so that those who want to use the facilities can give a visual signal that they are waiting. Finally, place grounded

Sinks with the same design but different heights work equally well for wheelchair and standing users. Storage cabinets are designed around the homeowners' storage needs. Double cabinet doors reflected in the mirror to the right open to expose a full-length mirror, with the toilet behind it.

A wall-hung sink with base cabinets on each side provides ample countertop space and storage for toiletries, linens, medicines, and first-aid supplies. A cabinet beside the toilet puts extra tissue within reach.

electrical outlets where they can easily be used for small appliances such as electric toothbrushes, hair dryers, shavers, and the like. The apron panel below the vanity is a good place for outlets, as they can be reached without stretching across the counter or having the electric cord come into contact with the water.

After leaving the bathroom we usually go to the bedroom, sleeping and dressing areas that are integral with bathrooms in the accessible home. In close proximity, both are intimate spaces apart from shared living areas. What do these rooms have in common, aside from being where we start and end each day? We turn our attention next to spaces used as bedrooms, dressing areas, and closets. ✦

STAYING TOASTY

An accessible bathroom is a toasty bathroom. People with disabilities may take a little more time attending to personal care, and so creating a comfortable bathroom environment is a top priority. Infrared lamps and radiant ceiling panels, which warm humans rather than the air around us, are a good idea in a space where wet bare skin is exposed. In-floor radiant heating systems keep room temperatures uniformly warm and avoid the need for space-consuming radiators. Several types of heated towel warmers are available on the market, designed for compatibility with the home's heating system.

Custom-designed concrete basins enabled the designers to create a free-form bathroom, full of sensual curves, and tailored to the homeowners' dimensions.

CHAPTER 7

dressing and sleeping

SLEEP PLAYS A CRITICAL ROLE in immune function. It boosts the body's ability to ward off harmful environmental pathogens that cause disease and it speeds up the healing process. According to the Centers for Disease Control, improving people's sleep habits would reduce the incidence of chronic diseases such as diabetes, cardiovascular disease, hypertension, cancer, obesity, and depression, as well as improve the management of these conditions. It would avoid over 56,000 reported highway accidents each year and over one million medical errors.

Open shelving puts everything within sight and within reach. A pocket door at left leads to a bathroom and a swinging door at right leads to a shared hallway and laundry. The homeowner, who uses a wheelchair, prefers the swinging door when he needs to carry a lap-full of clothing fresh from the laundry.

"If I were asked to name the chief benefit of the house, I should say: the house shelters day-dreaming, the house protects the dreamer, the house allows one to dream in peace."

—GASTON BACHELARD, PHILOSOPHER

To be sure, the epidemic of a chronically drowsy population has much to do with social, economic, and medical factors outside the scope of this book. As architects and homeowners, however, we should be asking ourselves how the design of sleeping areas can make a difference in keeping the immune system strong so that people stay healthy. In this chapter we look at the places where we sleep, including the adjacent spaces where we keep our clothing—also known as the bedroom and closet areas.

CREATE OPPORTUNITIES FOR SOUND SLEEP

Ask most people what makes a great bedroom and they're likely to say it's the mattress or sheets. We know that design is a factor, however, because of what doesn't work—noise from adjoining rooms keeps us awake, for example. Too-small bedrooms that result in clothing strewn on the floor are also high on the list. Poor lighting—the sort that lets our partner read in bed while making it hard for us to sleep—is another culprit. These factors are all amenable to design intervention.

> FOR MORE ON THIS HOUSE, SEE **Case Study 10 (p. 146)**

By placing master bedroom doors leading to the deck and bathroom far apart from each other, the path of travel to each doubles as transfer space onto the bed. High ceilings add visual interest for a person lying down.

QUIET

Provide acoustically absorptive surfaces such as carpets and window coverings to reduce unwanted sound, both internally and from adjacent parts of the home. Renovations offer a good opportunity for acoustic upgrades such as de-coupling walls and ceilings from framing lumber (see the sidebar on p. 39) or replacing windows to add another layer of glazing. By locating closets between bedrooms, or between the bedroom and a noisy street, the walls and clothing also provide acoustic benefit.

DARKNESS

The ability to regulate natural light is critical in making a room conducive to sleeping well. If you're planning a new home, consider locating the sleeping areas on the east side to help calibrate the body's circadian rhythms (internal clock) to nature. On the other hand, if you have trouble sleeping you may want to have the bedroom facing west where mornings are darker. In a renovation you can easily adapt existing windows to reduce incoming sunlight (see the sidebar below).

WINDOW COVERINGS

Window coverings regulate daylight to block glare from the sun, and they also help control indoor air temperatures by reducing air flow through the glass. Window coverings come in a variety of styles—shades, curtains, draperies, shutters, vertical and horizontal blinds—and the choices can seem daunting. Knowing your functional requirements makes styling decisions a little easier.

Whether you want to control natural light, ensure privacy, or conserve energy, there are window coverings to suit your needs. For example, to block sunlight streaming in from above you want horizontal lines (blinds or shades), but if the sun beams in from the sides you want verticals. Honeycomb shades fill the window opening for privacy, but double-pleated models conserve more energy by adding extra layers of air space, for insulating value. Opaque materials block more light, and translucent materials disperse it. Each material and style of window covering has its own purpose and so the goal is to select materials, type, and controls to match your requirements; color and pattern are aesthetic and secondary decisions.

Operating systems come in manual and motorized versions, which have the advantage of removing cords that can pose a tripping or tangling hazard. Window coverings can be controlled either individually or in groupings, depending on how sunlight travels through the room. To have some windows covered and others exposed, go with individual controls, operated by either remote or by wall switches. For hard-to-

Multiple window coverings control heat loss and daylight to make a comfortable indoor environment.

reach windows or with dexterity limitations, choose motorized controls.

Any time you control the amount of daylight that enters the home, you also control the amount of glare on interior surfaces and reduce the amount of cooling needed to keep indoor air temperatures comfortable. Construct wide roof overhangs or sunshields on the exterior. Plant deciduous trees that block summer sun when the leaves are full and let in warming winter sunlight.

PRIVACY

Privacy, like daylight, can be controlled architecturally with window and furniture placement. To balance privacy with a sense of control over who may be entering the space, locate the bed with a view of the doorway and use screening elements such as built-in cabinetry or partial walls to separate sleeping areas from adjacent activity centers. Consider skylights and high windows for daylight without giving neighbors a view inside.

LIGHTING

An approach to artificial illumination that works well in other areas of the house—ceiling fixtures—is seldom right for the bedroom. For reading in bed, consider swing-arm fixtures or sconces mounted on the wall beside the headboard, or lamps set onto nightstands; you want fixtures that do not shine in the eyes of someone who is trying to sleep. Gently light the path of travel to critical areas such as the bathroom and egress (exit). Bedrooms benefit from mood lighting, so place fixtures on dimmer switches. For added convenience, make sure any lighting controls can be operated from beside the bed, as well as upon entering the sleeping area.

SIMPLICITY

Isolate areas of stimulating activities, such as a home office, from the sleeping area by using screening techniques such as room dividers, cabinets, furniture, changes in flooring or ceilings, or changes in lighting. Repurpose a busy bedroom as a place for relaxation by making space for passive recreation—for example, comfortable chairs for reading or audio/visual equipment for listening to music. Plan the space to make it easy to implement bedtime routines.

Exercise equipment in the master bedroom allows the owner of this house to pursue her fitness routine at the start and end of each day. By keeping a strong upper body, she can maintain her independence and continue to use a manual wheelchair rather than a power model.

SERENITY

Simplify wall surfaces and avoid elaborate wallpaper designs to keep the setting soothing. Use artwork and paint colors to promote a restful environment (see the sidebar on p. 76). Where possible, design a ceiling that is worthy of the user's attention; cathedral ceilings, skylights, and vaulted ceilings with cove lights reward the hours spent looking up with a pleasing view.

CLEAN AIR

Use hypo-allergenic materials and finishes, products less likely to cause an allergic reaction. In a space where we spend about a third of our time, the benefits of breathing clean air for a large chunk of time can have lasting effects. If you are prone to allergies, defining what constitutes a clean room may take trial and error (see Case Study 11, House for Clean Living, p. 152). In general, use natural and nontoxic materials, biocide-free products, and low-VOC (volatile organic compounds) paints in neutral

THE PSYCHOLOGY OF COLOR

Colors have a strong effect on our moods and behaviors, and many believe these effects are universal. Color therapists see color as a vibrational healing modality, with each color having a unique energy that balances the energy in our bodies. Many designers of healthcare environments apply the same principles in choosing paint colors. Here are some of the qualities associated with colors:

- Yellow: cheery
- Orange: energetic
- Red: stimulating, comforting
- Purple: spiritual, exotic, wise
- Blue: serene, ordered, sad
- Green: natural, healthy, fertile, tranquil
- White: pure, innocent, sterile, bland
- Brown: reliable, conventional, down-to-earth
- Black: powerful, mournful, life and rebirth

This list suggests we should all have yellow kitchens and blue bedrooms, but personal experiences and cultural traditions also influence choice. Think about how colors make you feel when decorating. What colors are energizing? Soothing? Are there colors that make you happier? If you like bold color but find it too intense, consider painting an accent wall or bringing the color into your furnishings. Color choices are too important, and too individual, to be driven by the latest design trend.

Adding a bay window is a good way to gain space for transfer without enlarging the full bedroom.

colors. And, of course, make sure you have a great mattress with pillows and linens to your liking.

DESIGN SAFE AND SECURE BEDROOMS

Design the bedroom layout around transfer logistics as well as furniture requirements. Does someone in the family use a transfer board? Include a place to store it nearby. Can the person change position in bed independently? Design solid supports such as a headboard, nightstand, or grab-bars. Would an overhead lift connecting the bed and bathroom areas promote independence? Install solid blocking above the ceiling and raise the bathroom door opening for a continuous track (see Case Study 18, Modernist Chicken Coop, p. 185). If space permits, design a custom bed stand at the proper transfer height and with a wide transfer bench built into one side.

Create a generous path of travel connecting the bed with related spaces such as the bathroom and clothes closet. Make sure there is a 30-in. by 48-in. clear approach space on both sides of the bed and a 5-ft. clear floor area for turning; even if you don't have a wheelchair, the accessible home makes it easy to add assistive devices in the future. In designing the room, locate windows and doors to control daylight and privacy.

FIRE SAFETY

Planning around fire safety eliminates a huge source of stress for a person with disabilities. Codes require all sleeping areas to have a direct way out in an emergency—usually a window with stated opening clearance located within reach of the floor. For a person with physical disability, an egress door is preferable, leading to outside the home or to an area of "refuge assistance," a safe and

A sturdy metal-latticed headboard helps the homeowner shift position in bed, as he can grab it to turn his body during transfer or to reach the alarm clock.

fire-resistant space to wait for help to arrive. Construct walls and floor-ceiling assemblies around the sleeping area using fire-rated building materials, so that if there is a fire on one side of the wall, it takes longer to reach the bedroom, thereby giving the occupant time to escape. Choose fire, smoke, and intrusion alarms that use audio-visual (flashing lights alongside buzzers) or verbal annunciation devices.

DE-CLUTTER

Bedrooms can be clutter magnets. The accessible home takes control of how and where things are tossed so that it's safe to move about. Make sure the bedroom has ample and convenient storage, including the following:

- Shelf space for essential items beside the bed—alarm clock, watch, glasses, reading materials, tissues, phone, remotes, water bottles, and so on.
- Hooks for transitional clothing such as pajamas and bathrobes.
- Hampers for dirty clothing.
- A wardrobe stand or bench to lay out clothing for the next day.

- A designated parking area near the bed for mobility and assistive devices.

ASSISTIVE TECHNOLOGIES

For a person with breathing difficulties, the bed area may hold oxygen tanks and tubing. For a person with dexterity limitations or for anyone seeking added convenience, most home-electronics devices can be purchased with remote controls: a ceiling fan, window coverings, television, a sound system, an intercom, or a web-cam linked to the doorbell, for example.

For a person with hearing loss, the sleeping area is a natural place for locating a dry-store for sanitizing hearing aids and charging batteries for hearing devices, such as FM receivers, so they are ready to go each morning. Other devices for the hard-of-hearing include amplified and captioned telephones that light up when a call is coming in and alarm clocks that vibrate the bed. All these devices use power, and a sleeping room with ample

Doors from the master bedroom serve as emergency egress and also, with a deep roof overhang, a gentle connection to the outdoors.

HEATING, VENTILATING, AND AIR CONDITIONING (HVAC) SYSTEMS

A home's HVAC system regulates indoor climate. Current energy codes and a growing awareness of the environmental effects of fossil fuels are changing the way the building industry designs homes, making effective and efficient HVAC systems more important than ever. As building systems represent a major investment in home infrastructure as well as performance, the expertise of a qualified mechanical designer is advised. Distribution (piping and ductwork branching out to living areas from central equipment) needs to be closely coordinated with the architectural design to maximize usable interior space.

A well-designed HVAC system is tailored to both regional climates and the homeowner's needs. When the homeowner has hearing and visual limitations, noise control is an important factor, and impacts the decisions on using a ducted (air-based) vs. a piped (water-based) system, as well as where and how equipment is installed. When the homeowner is unable to self-regulate body temperatures, indoor air temperatures and humidity levels are critical factors in creating a comfortable environment; moist air feels warmer and saves on heating costs, yet too much moisture promotes mold growth. When the homeowner has environmental sensitivities, effective air filtration systems are most important.

Controls are as important as equipment and distribution, and affect comfort, safety, ease of use, and operating costs. Many new systems employ digital controls, which can provide many more functions than conventional thermostats. Choose easily legible controls and mount these within view. Hand-held remotes can be operated from anywhere to control temperature, fan speed, and air direction. Internet interface allows the system to adjust according to weather forecasts. With so many new technological capabilities it is easy to be tempted into buying an elaborate system, but unless the system matches the homeowner's human capabilities it will not function properly. As with other details of the accessible home, tailor the installation to the homeowner's abilities, skills, and performance requirements.

electrical outlets, located in the area of use and at a convenient mounting height, is a cornerstone of the accessible home.

CHOREOGRAPH DRESSING ACTIVITY CENTERS

Closet planning has become a specialty within the home-improvement industry, and for good reason. More clothing can be fitted into a confined space when storage is well planned. In old houses closets were barely deep enough to hold a coat hanger, and homeowners had to reach deep into side recesses to find things. Widening the closet and doors has been an evolutionary milestone in creating useful closets, but the accessible home goes further.

CUSTOMIZED STORAGE

Clothing storage is not a one-size-fits-all matter. People who live in colder climates have to store clothes for four seasons. People who dress up for work have two wardrobes—casual and formal—whereas those who work where every day is "casual Friday" may choose to wear jeans all year. Like clothing itself, the best closets are tailored to fit the owner. Whether one uses a bureau and closet or built-in armoire, storage needs vary greatly, and customizing the storage provides the most economical use of space. Cabinet materials are $3/4$ in. thick—much thinner than typical $4 1/2$-in. closet walls—enlarging available space for clothing in the bedroom.

Some clothing is best stored in slide-out drawers: underwear, socks, casual pants, and shirts, for example. Open shelves are more appropriate for foldable clothes such as sweaters and men's laundered shirts. Hanging rods hold jackets, skirts, dress tops, and pants. Rods mounted at two levels work well for short items, but full-length clothing requires

Compact shelving and a well-lit countertop are flanked by closet poles at two levels to make efficient use of space in this his/hers closet. Constructed as a built-in armoire, the entire cabinet area can be removed to connect two adjacent bedrooms into one larger space.

a single high rod. Shoes need racks or cubbies. Ties need racks and belts need hooks. Scarves and hats can go on shelves or in drawers.

CLOSET PLANNING

In planning a closet, there is no substitute for actually measuring the total linear footage of hanging items and the square footage of drawers and shelf space. Many homeowners have clothing stored all over the house, so the designer needs to hunt it all down to include everything in the planning. Consider how off-season clothing is stored, and whether a cedar closet or chest would be used. These areas may be located farther from the sleeping area but should be within convenient reach.

CLOSET DESIGN

Whether in the kitchen, family room, bathroom, or sleeping area, accessible cabinetry has similar hardware requirements (see the sidebar on p. 56). Use easy-glide drawers and large C- or D-type pulls rather than knobs. Locate closet rods within the user's reach range, usually less than 48 in. above the floor for someone who uses a wheelchair. Design bench seats, where used, at 17 in. to

With adjustable shelving, a clothes hamper, and coat-hanger poles at different levels, a closet in a converted barn is tailored to the height and reach of the homeowner.

His (at right) and hers (at left) accessible cabinets help keep peace by giving both homeowners adequate space for their own belongings.

> FOR MORE ON THIS HOUSE, SEE
> Case Study 6 (p. 125)

19 in. above the floor. Choose closet doors (pocket, sliding, bi-folds, or outswinging) to maximize usable space in the room, or omit doors altogether.

DESIGN DRESSING AREAS TO PROMOTE INDEPENDENCE

Knowing how a person gets dressed allows the designer to plan a space that works. Install hooks for hanging the day's clothes. Install grab-bars useful for self-assisted standing. Provide space for turning about in a wheelchair or for having a caregiver assist with getting dressed. Consider a bench for laying out clothing or putting on shoes.

A vanity is as welcome in the dressing area as it the bathroom. Provide places for putting on jewelry and make-up, and for hair care, with appropriate small appliances. Make sure there is adequate space under the counter for knees and a wheelchair footrest (see the sidebar on p. 50). Locate the vanity for glare-free but adequate lighting—beside rather than in front of a window, and with strong and controlled lighting.

Include mirrors in the plans as space permits. At least one should be full height, another at eye level if a vanity is installed. Locate another mirror on a swinging door or swing arm, for checking the back before heading out for the day. Make sure there is adequate lighting too in the dressing area to reduce the chances of mismatching socks.

Closet areas have a close affinity with the laundry room, one of the last stops on our virtual tour of the accessible home. We turn our attention next to the utility areas, which also include the garage and mudroom. ✦

An accessible vanity between the closet at right and the bathroom at left makes a comfortable place for applying make-up and jewelry.

This dressing area is conveniently located between the master bedroom (left), the bathroom (beyond), the closet (far right), and the home office (near right). A tall mirror and overhead windows ensure even illumination and easy viewing. A freestanding armoire separates the bedroom from the dressing area for visual privacy.

CHAPTER 8

utility spaces

MUDROOMS, LAUNDRY ROOMS, and garages: three messy rooms, workplaces to be blocked off with closed doors when guests arrive. Whatever our standards for home care may be, utility rooms are places where it is tempting to let go into the chaos. But in the accessible home, where clutter equals hazard, these rooms offer an opportunity to create little pockets of calm.

A compact laundry fits into a small closet between the bathroom and hallway. Wire racks and a built-in counter put essential tasks within reach. Bi-fold doors with 360-degree hinges lie flat against the wall in the open position. Staple-type pulls centered on the door panels make these doors easy to open and shut.

DESIGN LOW-MAINTENANCE MUDROOMS

The smaller the room, the more critical it is to plan space carefully, and mudrooms are no exception. As the link between indoor and outdoor circulation patterns, mudrooms are transitional spaces but also activity centers in their own right. Mudrooms do not need to be large—a short length of wall with hooks is enough—as long as the path of travel and storage are both accessible. Regardless of its size or contents, the mudroom must be large enough to maneuver in safely, free of swinging doors that intrude on usable floor space. This means a 36-in.-wide path of travel and a 30-in. by 48-in. space beyond the arc of the door swing and in front of storage areas. A vestibule with inward-swinging doors on opposite walls should be at least 7 ft. deep, measured from the door edge in its open position to the opposite door face. In taking a look at available space in your home, you may see room for expansion into an adjacent room or porch to right-size the mudroom. To design the mudroom, start by taking an inventory of your storage needs and other room requirements:

- How many people arrive at once? Make sure there is space for people to change out of outdoor clothing.
- How many people leave at once? Create a place where lunchboxes and gym bags can be grabbed as the family dashes out of the home each morning.

> FOR MORE ON THIS HOUSE, SEE **Case Study 15 (p. 168)**

- Is the mail drop located in the mudroom? Provide a container so that mail does not clutter the floor and is easy to retrieve without bending down.
- Is your home a shoe-free zone? Design cubbies for changing out of footwear and into slippers.
- Does your family play sports? Lockers and drawers offer two approaches to stashing gear.
- What do you want to store? Keys, backpacks and handbags, boots and shoes, outdoor clothing—each item belongs someplace.
- Do you recycle clothing when seasons change? Coat hooks are fine for the

Dark colors make the mudroom easy to keep clean. The room contains a charging station for the homeowner's power wheelchair.

Open shelving beside the front entrance provides a handy place to put things when entering or leaving the house. The top surface is sized within reach of the person using it.

Mudroom cabinets at two levels place coat hangers and drawers within reach, whether seated or standing.

BENCH DESIGN

Design benches at the same height as a wheelchair seat, usually 17 in. to 19 in. high. If back rests are provided, place them 18 in. high above the bench, and not more than 2 in. above, or 2½ in. behind, the seat. A bench seat should be at least 42 in. long and from 20 in. to 24 in. deep, designed to support a load of 250 lb. at any point. The seat surface should be slip-resistant (and, if outside the house, designed to prevent water from accumulating). Consider adding grab-bars on a wall beside the bench, but not behind it.

short term, but hangers prevent clothing from stretching when stored for any length of time.

Count the items you need to store, and measure the space requirements for each grouping. Measure also the reach ranges of all residents so that stored items are easy to get at. With a list of storage items in hand, consider how to optimize space in your mudroom. Hooks mounted above head height max out wall storage over a bench. Drawers offer efficient storage close to the floor, and are particularly well-suited for small items such as mittens and hats. Shelves put everything within view, and are the best way to store items high on the wall. Whether or not to enclose shelves is a personal choice: If you want ready access but can tolerate visual clutter, omit the cabinet doors and go with open shelving. Don't let any space go to waste—the area beneath a bench can be used for boots or filled in with drawers stowing sports gear.

If someone requires assisted care, consider creating a place where mobility devices can be cleaned and stored. For people who use a scooter outside and a walker inside, or who have both power and manual wheelchairs, size the mudroom for transfer logistics as well as parking. Battery-charging stations make sense where power devices are parked. Provide space for two power chairs—one for use while the other is charging—a handrail for self-assisted transfer, and electrical hook-ups. Provide wall hooks for navigational canes to show consideration for guests with low vision.

Choose durable and low-maintenance materials for walls and flooring. Consider mudroom acoustics, as this can be a noisy space (see the sidebar on p. 39). Paneled walls are more impact-resistant than plaster, and impact is inevitable in any active room where space is at a premium. Wall paneling presents an opportunity to integrate handrails into the design. Consider inset floor mats so that sandy or wet floors do not present a slipping hazard. And as with other areas in the accessible home, uniform and glare-free lighting helps ensure safety while using each space.

> *"A well-planned closet stretches the home's storage capacity by up to 60 percent over a conventional coat-rod and shelf."*
>
> —PAM SMITH, CLOSETMAID

Locating a laundry on the main hallway connecting bedrooms makes it easy for children to learn skills of independence and self-care, while the wide hall doubles as a staging area. Shelves tucked into the side wall max out storage, and adjustable wire racks allow flexibility in use.

DESIGN CONVENIENT LAUNDRY AREAS

For people with disabilities, laundry activities offer a way to stay engaged with housekeeping. Locating the laundry near the center of the home shortens travel distances, an important quality for those who have difficult moving around. Any place you can bring plumbing and find room for a washer and dryer is a potential site for the laundry. Consider where the family spends its time. A laundry near bedrooms or living spaces will make it easier to pop damp clothing in the dryer while going about the day's activities.

Whether to purchase top-loading or front-loading equipment is both a matter of personal preference and the user's mobility requirements. Appliances can be stacked to take up less space, although

LAUNDRY CHUTE
If you need to put
the laundry on a
lower floor from the
bedrooms, consider
installing a chute,
with the opening
both small and high
enough to prevent
people falling
through, yet low
enough to reach
easily inside.

the top machine becomes harder to reach. Side-by-side machines designed for front-loading are more convenient for a seated user, and many people construct platforms to raise appliance doors put the contents within view. Appliance doors should be no more than 36 in. high for top-loading washers, or from 15 in. to 36 in. high for front-loaders.

As with kitchen appliances, controls for washing machines and dryers come in a range of options to suit people with a variety of disabilities. For those with low vision, rotary dials with pointer indicators that click audibly into distinct positions allow the user to track the wash cycle by feeling the arrow location. Push-buttons with audible tones are made more user-friendly when the button surfaces are embossed. Beware of endlessly turning dials or flat-screen touch panels with electronic visual displays; these can be hard to use. For improved visibility, choose appliance controls with large

bold lettering, contrasted strongly with background colors, and display screens that produce less glare.

A well-equipped accessible laundry has shelves for detergent, bleach, fabric softener, and measuring cups. There is a counter for folding clothes and stacking clean laundry, and a sink for hand-washing delicate fabrics. Drying racks or a clothesline pull out from the wall. There is a nearby ironing station, with a fold-out wall-mounted ironing board, iron, and space for hanging freshly pressed clothes. As with other accessible activity centers, light-fixture locations are coordinated with the machine door openings and work areas.

"Lugging a vacuum cleaner around the house isn't easy, so install an in-wall vacuum system instead."

—BLIND HOMEOWNER

OPTIMIZE GARAGE SPACE FOR STORAGE

Garages, basements, and utility closets are places where stuff accumulates, sometimes falling on the floor and often making it difficult to find what you want. The same guidelines that allow us to create accessible mudrooms and laundry areas apply to other utility rooms. Sufficient shelving, hooks, drawers, and hangers—tailored to the items stored—results in utility rooms that are safe and convenient. Sturdy finishes simplify maintenance, and effective task lighting reduces the chance of accidents. Providing ample electrical outlets near areas of use keeps extension cords above the floor. By coordinating wall storage with the user's path of travel we keep maneuvering spaces wide.

The kinds of interventions described here are not difficult to make. Home

This well-stocked laundry room has a generous work counter for folding clothes and a sink for hand-washables. Vases displayed on open shelves attest to the room's other function—as a place for arranging flowers fresh from the nearby garden.

PLAN FOR PETS

Guide dogs, watchdogs, service dogs, or just companions, pets are part of the household in many accessible homes. The designer can make pet care easier by planning places for cleaning and feeding the furry housemates. The mudroom becomes a pet station with a little planning. Is cat litter or pet food stored here? Provide appropriate containers to keep the floor from becoming slippery.

A room designed for wet household functions—a mudroom, laundry, or shower room—has an affinity with other messy household operations such as washing and grooming. Depending on Fido's size, install a hose bibb or a utility sink with a gooseneck faucet. Choose nonslip flooring and install a floor drain, with a trap primer to keep fumes out of the home (see p. 66). As with all activity centers, adequate storage for supplies is a must: in this case, doggie shampoos, clippers, scissors, towels, and brushes.

Two large watchdogs are family members as well as part of the home's security system. This accessible home has places for pet care—feeding, washing, grooming.

A base cabinet in the laundry room is fitted up as a cat litter box. The cat enters the round hole, and the owner slides the drawer forward to scoop out the litter box.

organization systems are readily available at hardware stores and online, and an industry of consultants has sprung up to help people make better use of their basements, garages, closets, and kitchens. Stepping outside the accessible home now, we find that questions of storage become secondary to the task of creating living spaces to serve our outdoor lifestyles. ✦

A well-organized storage wall in the garage keeps floor areas clear of tripping hazards.

CHAPTER 9

outdoor places

YARDS ARE THE OUTDOOR equivalent of indoor living areas, and with a little planning, the yard can have something for everyone. Whether disability occurs suddenly or gradually, once-pleasurable outdoor activities such as walking or tossing the dog a ball may seem forever lost. But by linking outdoor activity areas on a safe and sturdy path of travel, the accessible home puts the natural world within reach again.

Think of how you live when planning your yard. Do you host the gang for outdoor barbecues, or do you prefer quiet mornings with tea and the newspaper?

An open porch near ground level makes a sunny spot to sit for much of the day. Extending the porch out into the yard gives views in several directions and increases natural ventilation.

"What if every landscape...was conceived equally from inside and outside, in a dynamic relation to its surroundings, as a living, breathing organism that has a one-of-a-kind personality?"

—TAMARA ROY, ARCHITECT

Do you play catch with the kids or tend a garden? Whether you enjoy active or passive recreation, there is an accessible yard for you. Accessible toddler play structures, outdoor sports, picnic tables, bird feeders, and places from which to observe all of the above: The accessible home has these features and more. By purposely creating new ways to enjoy the outdoors, the accessible home enriches daily life for homeowners and visitors alike. Let's look at some of the ways this can be accomplished.

BRING NATURE INTO THE HOME WITH INDOOR–OUTDOOR ROOMS

Imagine a continuum of accessible living spaces between indoors and out, from enclosure to exposure, and you can see the possibilities for blending interior and exterior environments. We start with a typical indoor room, defined by walls and punctuated by windows—say, a living area. Adding glass gives the room a stronger link with nature; the space becomes a sunroom. When the windows are replaced with screens, we have a screened porch. Remove the screens and it is an open porch. Remove the roof and it is a deck. Lower the deck to ground level for a patio. Remove the paving and you have a yard. Take away the lawn and you're in the forest—or beach, meadow, prairie, desert, and so on. Many of the homes featured in this book have a variety of ways to experience the outdoors, all designed in ways that reflect the architectural

character of the house as well as the lifestyles of homeowners. For a person living with disability, indoor–outdoor activity areas bring the natural world and all its benefits within reach.

SCREENED PORCHES

Screened porches invite the outdoors inside simply and easily, without forcing a change in mobility devices, or even a change in clothing in many climates. For a person with limited senses, outdoor smells and sensations become anchors,

A paved driveway makes a natural accessible play yard when the garage is located close to the house.

OUTDOOR LIVING FOR HEALTH AND HAPPINESS

A growing body of evidence points to the positive health benefits of engaging with nature. Direct contact with nature promotes psychological well-being, reduces stress, decreases fatigue, restores mental clarity, improves hardiness and a commitment to self, and aids in physical healing. The good news is that all kinds of nature experiences confer these benefits—from gardening to wilderness backpacking to watching clouds drift by overhead.

A screened porch projecting outward from the lines of the house is one way to live surrounded by nature. Storm windows replace screens during cold weather to create a three-season room.

signaling the time of day or season. Run the screens from floor to ceiling for natural ventilation and a view to the yard. Install skylights and screened transom openings for star gazing. Locate framing elements to balance unobstructed vistas with the safety of a sturdy guardrail system. Maintain flooring levels and even illumination levels to make the screened porch an accessible extension of the home's living areas.

OPEN PORCHES

When the yard is inaccessible, porches of all types provide another way to connect with the outdoors. With a roof and a deck but no walls, the open porch can be simply an extension of the entrance landing or an outdoor room in its own right. An open porch offers weather protection at exterior doors—an important function when floors and decks are at the same level. Size a deep porch beside the kitchen for outdoor dining, or design a shallow wraparound porch for 360-degree views of the yard. Recess the porch for a protected outdoor sitting area, or extend it out into the yard to be surrounded by nature.

DECKS

A deck is a roofless porch, and an appropriate substitute where weather protection is not needed. Many decks are extensions of indoor living areas, transitional spaces linking rooms with yards, fitted with ramps and/or stairs leading to grade. When the home's upper floors are accessible, roof decks take space that would otherwise be unusable—for

> FOR MORE ON THIS HOUSE, SEE Case Study 19 (p. 189)

example, above a garage—and convert it to a functional outdoor living area. Detail deck edges to prevent people from slipping off, following code requirements for handrails or curbs, depending on the deck height.

BALCONIES

Consider adding a balcony to gain a little more floor space and a view to the ground below. Even a tiny Romeo and Juliet balcony allows you to add a full-length door in place of a window for greater connection with nature, and at the same time, to introduce more daylight into the home. Whether they are designed to project out from the building face or recessed into rooflines, balconies maintain a building's footprint while gaining outdoor living space.

CREATE OUTDOOR ACTIVITY CENTERS

A freestanding structure may offer a way to gain accessible outdoor living space if the house itself cannot be made larger. One couple who designed their home for aging in place has a Thai-inspired "sala," a freestanding room with screened walls, reachable from the house via an open deck (see Case Study 23, Aging in Place without Adding Space, p. 213). In another community a man with MS and the homeowner next door commissioned a covered outdoor bench with seats facing the two adjoining yards—a mutual amenity rather than a barrier between neighbors (see Case Study 24, Accessible Tower, p. 216). For a person with limited mobility, having a protected place to rest on a journey through the yard makes the outdoors more user-friendly.

> FOR MORE ON THIS HOUSE, SEE Case Study 9 (p. 140)

This accessible roof deck over the garage ensures that the second-floor level has generous outdoor space. As the shadowline indicates, the roof is fitted with slats designed to block the low western sunlight from overheating the home.

> FOR MORE ON THIS HOUSE, SEE Case Study 10 (p. 146)

A small deck at floor level off the master bedroom provides an accessible private outdoor space.

CHECK WITH
ZONING
Porches, decks,
and balconies
offer exciting
opportunities for
outdoor living,
but there may be
limits to what can
be built under local
zoning regulations.
Check your town's
requirements, as
existing houses may
have the buildable
area already
maxed-out. It may
be worthwhile to
apply for a zoning
variance if this is
the only way to gain
accessible outdoor
space. For detached
structures, town
laws may be more
lenient.

This outdoor room—or sala—is linked to the house by accessible decks and ramps, both indoors and out. Wide walkways, benches, and railings help to make the path of travel safe.

OUTDOOR STRUCTURES

Sometimes all a yard needs is a little shading to control daylight. For a person unable to regulate body temperatures, or who finds high visual contrast distracting, a freestanding outdoor structure can be a welcome addition to a sunny pathway or deck. Pergolas define space without actually enclosing it, using slats sized and angled according to the sun's orientation to reduce glare and improve visibility. Climbing vines block even more sunlight. For someone who may be directionally challenged, the sight of a pergola hovering over a portion of the yard is a beacon that signals the way home.

OUTDOOR RECREATION

Playground equipment has come a long way since the introduction of the ADA. Modular accessible outdoor structures include swings, see-saws, slides, ladders, bridges, and climbing structures that promote large- and small-motor activity for children of all ages and sizes. Sports have also come a long way, with Paralympics and Special Olympics games providing new inspiration to stay active.

Outdoor pathways are defined by site structures. The fence separates landscaped gardens, designed in part for the public's enjoyment, from the family's private yard. The long pergola shades an otherwise sunny pathway.

One homeowner who uses a wheelchair had an outdoor swimming pool installed in her yard for daily upper-body strength training (see Case Study 6, Low-Effort Living, p. 125). Another homeowner who is blind had a cable installed alongside a pathway on his lot so that he could navigate using a white cane. One yard in this book is packed with places for entertaining friends and mother–son sports, including a sandbox, basketball hoop and horseshoe posts, hot tub, vegetable and flower gardens, and barbecue grilling and picnic areas (see Case Study 17, A House that Grows with Its Owner, p. 180). In designing accessible outdoor places, plan wide pathways with gentle grades, ramps, and handrails for safe travel (see chapter 2). Consider the comfort and safety of both the outdoor athletes as well as the caregivers and observers.

CHOOSE OUTDOOR FURNISHINGS FOR DURABILITY AND COMFORT

Take the same care in furnishing your porch, deck, or yard as you use in the home. Choose comfortable chairs and benches to match the homeowner's mobility devices and transfer logistics. Install built-in furnishings to keep the path of travel clear for a person with low vision or assistive equipment. Other site features may include freestanding mailboxes, bird feeders, birdbaths, and parking areas for mobility equipment. In the accessible home, each of these features is on an accessible path of travel and has an adjacent 30-in. by 48-in. space for a wheelchair.

OUTDOOR DINING

Finding the right spot for outdoor dining requires that we choreograph an accessible path of travel for both people

Subtle ramps integrated with the site grading puts the owner's swimming pool within reach. The gently sloped ramp and stairs provide two ways to reach the swimming pool as well as lawn areas. A simple sun shelter was sized and located to provide poolside shading. Artificial turf makes a durable year-round lawn.

A cable strung between posts marks a path through the forest for an outdoorsman without sight.

Choose pavement materials that are firm, slip-resistant, and sloped slightly so that water does not accumulate on walkways, and provide handrails where needed for stability.

Paving made from recycled rubber tires makes a surface soft enough for falling, but also firm enough for using a tricycle or wheelchair.

This backyard has an accessible deck large enough for the extended family. Raised bed planters to the left keep the family supplied with fresh grown vegetables for much of the year, and are located at an accessible height where the whole family can help garden.

and food. Choose picnic tables with space for wheelchair seating, and avoid those with built-in benches. Set the gas grill into a freestanding island with knee space below. If you prefer to cook indoors, construct a pass-through from the kitchen to reduce the distance between preparing and serving meals. Create an accessible walkway to the street for guests. Include trash disposal in the planning. Whenever the path of travel can be shortened, the home becomes more accessible.

FENCES

If your household includes a family member who is prone to wandering, you will want some kind of enclosure. Paved or landscaped, fenced or walled in, a sheltered yard combines the safety of the built environment with the sense of adventure that accompanies a foray into the natural world. Semi-enclosed spaces between the house and garage, or where an addition joins the original house, morph easily into courtyards. Soften yard edges using curved fences rather than 90-degree turns for a person with cognitive impairment who uses a fence to orient himself in a large area. In designing a fence, consider the experience of those on the opposite side, as well as the architectural character of the house. The best fences contribute to the property value while being good neighbors.

DESIGN ACCESSIBLE LANDSCAPES

Lying on the grass watching the stars at night was a childhood pleasure for one homeowner, who incorporated a small lawn area beside a wheelchair transfer bench into her new yard (see Case Study 17, A House that Grows with Its Owner, p. 180). Another used artificial turf over gently graded paving for year-round greenery with zero maintenance, and a uniform surface for wheeled mobility devices. Yet another had a porous planting grid installed beneath the grass for a durable mud-free lawn. These are a few of the ways that accessible lawns can be created. Whether linked by paved pathways weaving through the yard, or accessible in their own right, lawns help keep the air fresh and cool for healthy outdoor living.

ACCESSIBLE GARDENS

There are so many ways to be a gardener these days, and an accessible garden is a wonder of ergonomics—good news for people with disabilities, backaches, and weak knees! Put the soil within reach by designing raised-bed planters, either in table-top boxes with knee space below them, or in solid boxes built up from the ground and designed for parallel approach in a wheelchair. Use window boxes for small plants such as a kitchen garden. Incorporate vertical gardens into a retaining wall or as part of a privacy screen. Smaller container gardens are easier to water and to weed than large gardens and make the joys of working with plants accessible to everyone. As with any garden, be sure you have a source of water nearby, and a handy place to store yard tools.

Looking closely at activity centers both inside and outside the home, we gain a new image of the home. Rather than have walls and doors define activity areas, the accessible home is open, more like a loft on the interior. Bedrooms and bathrooms are linked with pocket doors and wide hallways to create a spacious private zone within the home. Kitchens and utility spaces are designed like the cockpit of an airplane: tightly coordinated workspaces, where everything needed is within easy reach. Living areas inside the home are coordinated with those on the outside by a seamless path of travel. Building systems—such as acoustics, lighting, and

environmental controls—make each living space comfortable and safe. High-quality materials and well-constructed details reduce maintenance costs and chores. The accessible home promotes independence for its residents and is a welcoming presence for visitors. In the rest of the book, we will see how these ideas translate into real homes for real people. ✦

CONTAINER GARDENING Grow plants in rolling containers that can be moved around to absorb sunlight or define different seating arrangements. Choose containers in colors that contrast with the pavement so they can be more easily seen by partially sighted gardeners and visitors.

This raised-bed planter was constructed with a thinner wall at one side, allowing this homeowner to get close to the earth in her wheelchair. Mother and son grow potatoes, herbs, and vegetables in this garden close by the kitchen and outdoor dining area.

CHAPTER 10

historic homes

IT SEEMS NATURAL TO ASSUME THAT CREATING A new kind of home means starting from scratch, but old houses often lend themselves to especially intriguing solutions. Solid building materials, well crafted, give the house a strong character that serves to make the accessibility improvements seem less dominant. Extra spaces such as alcoves and foyers are easily adapted for elevators—properly located and without sacrificing functional areas. Old houses are often sturdily built and can stand the abuse of a construction project. In fact, many old houses have already undergone generations of upgrades, as any renovator can attest.

If the house is a designated historic landmark you may have special approvals to negotiate but it is usually an effort worth making. An older house that has stood the test of time will last for many more generations, and its value increases as the house become more relevant to today's realities.

accessible prairie style: the wright touch

When Maynard and Katie Buehler began planning their San Francisco home in 1947 they felt drawn to the principles in Frank Lloyd Wright's work. The house he designed for them in the Bay Area hills contains many of his signature elements: a large living–dining area and a private bedroom wing, with the house contoured to fit the landscape and ending in a carport. There was also an attached machine shop for Maynard and an outdoor swimming pool with cabana for their daughters. Although the word "accessibility" had not yet entered the vocabulary of designers, in Wright's earth-hugging designs lie the roots of many contemporary and accessible homes described in this book. It is in this spirit that the story of the Buehler house is presented here.

In true Prairie style, the Buehler house is characterized by strong horizontal lines and rooms that hug the landscape, both in form and materials.

Carports were covered parking areas with walls for ventilation and unencumbered access. By gently sloping the pavement at both the driveway and sidewalk to a common drain, the design keeps the ground level and dry.

REBUILDING A 1940S DESIGN WITH 1990S CONVICTIONS

Fast forward almost 50 years to 1994, when a fire ravaged the Buehlers' house, destroying all but the living rooms and machine shop. Maynard and Katie turned to Walter Olds, who had been Wright's apprentice architect when their house was originally designed. In addition to reconstructing burnt-out areas, they wanted improvements that would make this a good home for their later years, with a larger kitchen and a dressing room for Katie. Their grown daughters had left home and the swimming pool had been converted to a koi pond, the cabana to a guest house. It was time for a radical overhaul to the house.

Floor Plan

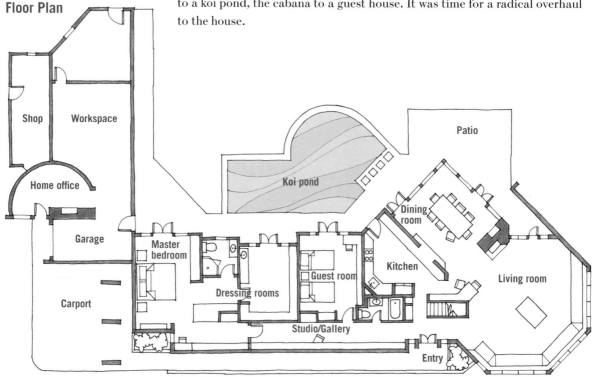

WHAT MAKES WRIGHT RIGHT

Inspired by the flat expansive calm of the Midwest prairie, Chicago architect Frank Lloyd Wright developed a unique style that revolutionized home design. In 1901, *The Ladies Home Journal* described a Wright house in its article entitled "A Home in a Prairie Town," and the name caught on. The Prairie Style house has low-walled terraces sequestering private gardens and wide sheltering roofs that hover above bands of windows. Living areas radiate out from a central fireplace, with level floors and glass-walled rooms that blur distinctions between indoors and out. Wright designed everything—walls, doors and windows, tiles and light fixtures, down to fixed and portable furnishings. Reacting to the mass-produced elements becoming popular in the building industry and the historic revivalism fashionable at the time, Wright's homes laid the groundwork for the contemporary movement in architecture and also for today's ranch houses.

Usonian Homes

In 1936, with the United States in the depths of an economic depression, Wright turned his attention to creating homes geared to the budgets of average Americans by eliminating attics and basements. Glass walls and deep roof overhangs allow passive solar heating and natural cooling to reduce utility costs. He designed open-walled parking shelters and coined the term "car port" to describe these. Using native materials, artfully detailed and skillfully crafted, Wright sought to create an aesthetic based on functionality rather than ornamentation.

Grid lines inscribed in concrete floors, horizontal wall paneling, and grooved-wood ceilings establish dimensional modules that create a sense of proportion and order, by defining locations for all built-in items—from door openings to shelf heights. He called his designs "Usonian," an acronym for the United States of North America.

Wide roof overhangs and a floor that flows outward to become a patio combine to create a covered accessible pathway around the house. Small skylights and in-set light fixtures ensure that the pathway is always well-lit.

Neither Walter nor the Buehlers are alive today, so their conversations cannot be reconstructed, and design intent can only be inferred by looking closely at the house and talking with people who knew them. Walter's wife was actively involved in Berkeley politics, where an activist community had played a major role in the recent enactment of the Americans with Disabilities Act. She recalls that Walter was deeply committed to improving accessibility in all of his work, and that the Buehlers were beginning to have problems walking, although neither of them would admit it. Walter knew that accessible features enhance options for staying at home, and also make a welcome place for friends with disabilities to visit.

Walter completely redesigned the central portion of the house from kitchen to master bedroom, widening outside walls to give Katie her larger dressing room. Damaged surfaces were restored at the living–dining area and machine

Sturdy built-in benches alternating with small tables surround the living area, keeping floors open and clear, whether for walking or using mobility devices. Furniture that does not roll away makes a more reliable seating surface.

shop. Walter's details were true to the Usonian principles (see the sidebar on p. 99) under which Wright designed the original house. At the same time, the design shows a thoughtful response to the myriad issues of aging gracefully at home.

DESIGNING FOR COMFORT AND SAFETY

The house is designed for indoor climate control, as older bodies self-regulate temperatures less effectively. A new concrete slab with radiant heating stabilizes the interior climate, and incoming sunlight is controlled with wide roof overhangs and clerestory windows atop exterior walls. Skylights with integral light fixtures illuminate outdoor pathways to enhance safety by day and night. The old gravel driveway was re-paved with rough-aggregate concrete

A built-in table extends from the kitchen to connect living and dining areas for serving and informal meals. Dining and living areas are visually connected by a decorative wood ceiling and open sightlines, making it easier for people with low vision or hearing to stay in contact.

and re-graded to floor drains, keeping walking surfaces safely dry. Knowing that one of Maynard's greatest pleasures was feeding the koi, Walter widened the patio between pool and house, edging walkways in stone and insetting glass panels for watching the fish at a safe distance.

Sometimes we have to look twice to realize that clever details are also sensible ideas to support multiple goals. Living well in later years is not incompatible with good design—rather, it can inspire and challenge the architect to improve on old models. Walter added built-in features throughout the house, designed for easy living by an older couple:

- The original two bathrooms were rebuilt in a compact design that puts everything within reach for minimal effort.
- A built-in desk in the guest bedroom has a fold-down table, easily shut to increase maneuvering space.
- A narrow window has a hinged shutter facing the driveway, as monitoring visitors increases one's sense of safety at home.
- Built-in armoire cabinets at the master suite are dedicated to each person's wardrobe needs and dimensioned for easy reach.
- Open countertops above Maynard's wardrobe keep sightlines open to give residents visual continuity and allow daylight to reach all corners of the room.
- A 35-ft. gallery connects living and sleeping areas, lined with a long, low desk and open shelving.

A desk in the master bedroom has a narrow window with a hinged door to provide a view of anyone nearing the front entrance. A clear view of parking and walkways enhances the sense of safety and reduces surprises when visitors arrive.

Built-in armoires eliminate the need for freestanding furniture and keep aisles free of clutter. Easy-slide cabinet shelves and drawers customized to the homeowner's clothing combine to create an accessible dressing room. Clear sightlines and natural lighting enhance non-verbal communication.

In widening the central part of the house, Walter reduced the number of sleeping rooms from three to two. Otherwise, rooflines were kept intact and the basic design was true to the original. Fortunately for us all, the Buehler House was rebuilt in a way Wright would undoubtedly have endorsed. ✦

Bedrooms with exterior doors allow exit in an emergency. Concrete floors make a smooth, level, and durable surface for mobility devices and connect directly with outdoor patios. A low shelf beside the desk is hinged to lie flat against the desk and widen the path of travel.

accessible, adaptable, and historic

John and Ann had always planned to downsize when they reached their 50s. In fact, they had written a clause into their marital contract, promising themselves to leave their suburban house and move to the city after their children were grown. It had been a deliberate choice to create a more sustainable lifestyle—close to work and the cultural opportunities of an urban center, driving less and walking more. As they considered their needs for a home where

Accessible details underlie decisions throughout the kitchen. Shelves are open and low, while glass dish cabinets are conveniently placed between the snack bar and adjacent breakfast room. A wooden handrail in front of the sink is both a handy support and a towel bar. Appliances are shallow, and rolling base cabinets can be removed for seated countertop use.

Driveway grades were lowered to carve out space for an accessible entrance at basement level. Excavation around tree roots required special approvals from the town's arborist. The ground floor holds a guest suite, and family living areas are on floors 2 and 3.

they could spend their "next 50 years," a small close-knit community where people cared for each other was important, as were the conveniences of a town within walking distance—and, anticipating future mobility devices, within rolling distance. Their requirements for the house itself included space for a garage and workshop. Three years after they began house-hunting, John and Ann moved into their renovated Craftsman-style home with a new elevator on the outskirts of Washington, DC.

TIMING THE MOVE

The couple had seen clients and friends forced to move when a serious fall or illness made it unsafe to stay at home. As an architect, John Salmen understood that careful planning now would reduce the chances they would have the expenses and disruption of a renovation (or a move) when they were least prepared to manage it. John and Ann knew their window of opportunity was wide open; they had the energy to invest in a move and the time to invest in making their new house a home.

They looked for a run-down house in an established neighborhood, expecting to renovate. The house that beckoned was in a historic district with strict controls on building modifications (see the sidebar on p. 106).

Floor Plans

First floor

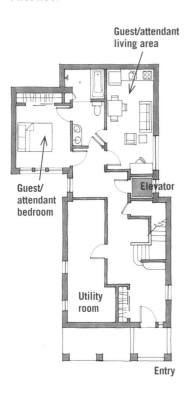

Guest/attendant living area

Guest/attendant bedroom

Elevator

Utility room

Entry

Second floor

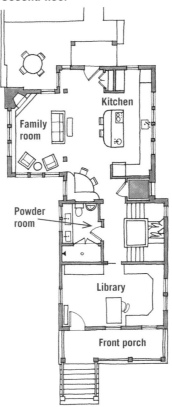

Kitchen

Family room

Powder room

Library

Front porch

Third floor

Balcony

Master bedroom

Studio

Structural steel beams (shown in blue at the ceiling) were used in framing to eliminate load-bearing walls and increase usable floor area at interior rooms.

An under-counter refrigerator and rolling trolley cabinet make good use of below-countertop space.

GAINING APPROVALS FOR UPGRADES

Two primary concerns in upgrading older houses for aging homeowners are access (getting around with mobility aids) and egress (getting out quickly in an emergency). When entrances in older buildings are modified, renovation plans may need approval by historic preservation commissions, whose members are charged with maintaining the area's original character. John and Ann's home was located in the largest historic district of the nation's capital, with a savvy and active preservation community. The first hurdle would be to plan the house's circulation patterns to ensure that the work would be approved.

Elevator placement would drive the layout, but the old roof was too low to accommodate the required overrun—usually about 12 ft. from the topmost floor served by the elevator to the underside of the hoistway (the vertical shaft holding the elevator cab). By designing an addition with an elevator at the back of the house, John could gain the requisite higher ceilings with minimal changes to the historic fabric of the house, and at the same time, bring some breathing space into the plan to accommodate future mobility devices. The new plan also gives each floor an emergency exit—from doorways to grade at the lower two levels to a balcony off the master bedroom at the topmost level.

Site modifications required approvals from the city arborist under an ordinance designed to protect older trees. Any proposed work within the drip line required an affidavit from a specialist to confirm that excavations would not damage the tree. For both the exterior house modifications and also site re-grading, working with the town to gain approvals ensured that the new work kept the neighborhood character intact, which is what drew John and Ann to this house in the first place.

A wood ramp connects the back door and garage, putting the main living areas at the second floor on an accessible path to the outdoors.

A three-story bungalow, the house had potential—and also a prominent front staircase and steep driveway. An accessible entrance would require lowering the driveway to cut a new doorway at basement level, and an addition to the house would change its appearance, so the project calendar needed to allow time for special approvals. When the good news arrived, the project went into high gear.

John designed an addition and substantial renovations at the house, in collaboration with historical and kitchen specialists. The addition contains a new kitchen and family room addition at the main level, with a master suite above and a guest suite below. Interior spaces at the old house were gutted and completely rebuilt. On the original house's first floor there is a library and accessible powder room. An art studio and laundry room occupy spaces formerly used as bedrooms on the second floor, and the basement holds a new main entrance and reconfigured mechanical room.

ACCESSIBILITY, ADAPTABILITY, AND FLEXIBILITY

John and Ann's house is Accessible. The gently sloped driveway puts the new front entrance a short distance above the sidewalk. A three-story elevator and wide hallways make for comfortable passage within the house, whether walking

Placing the new elevator beside the original stair allows the two means of vertical travel to share landings and provides equitable pathways for those with and without disabilities.

The first floor powder room has sinks at two heights and open storage beside the toilet. A wood grab bar along the countertop edge is used for assisted standing, as a toilet grab-bar, or as a towel rack.

Sliding cabinet doors at countertop level and a front-loading washer and dryer help make the laundry room accessible.

or using mobility devices. Halls and doorways are wide, countertops are installed at various levels, and the bathrooms have roll-in showers and extra space beside the toilets for wheelchair transfer or assistive help.

John and Ann's house is also easily Adaptable. Bathroom walls are framed with blocking, making it easy to add grab-bars at a later date. Lighting, electrical, and communication systems were integrated and selected for their ability to be upgradable, using compatible software packages and pre-wiring throughout the house. Modular cabinets in the kitchen and bathrooms can easily be removed for knee space or replaced with other types of storage as homeowner needs change.

And their house has features that are Flexible in use. A rolling cabinet doubles as kitchen storage and also a serving trolley. Hand-held showerheads on vertical rails can be easily adjusted for height. The basement office and family room layout converts easily to an apartment for a live-in caretaker, with plumbing and wiring for a kitchenette already installed within the walls.

These three qualities—Accessibility, Adaptability, and Flexibility—make the house a solid example of Universal Design in action. John and Ann know they will not be forced to move by a sudden change in their health or the economy. Sensible design removes many of the causes of falls at home, such as steps, tripping hazards at floor level, and reaching for high shelves. Good design makes the house a comfortable place to live, as well as welcoming to guests. These qualities combine to make the house a great home for John and Ann to live out their years together. ✦

urban loft

Most successful home modifications are designed around the homeowner's requirements, but when illness strikes, success is a moving target, elusive, unclear, and unpredictable. Randy and Leslie knew when they purchased their loft that a diagnosis of a rare hereditary disorder called Adrenomyeloneuropathy (AMN for short) meant Randy would eventually need help standing, walking, and getting around, although how soon this would happen, no one knew. And while they had no illusions that AMN was curable, they were hopeful that the right environment might bring some measure of healing.

Portable furnishings are easily rearranged to suit the size and type of gatherings, including community meetings and intimate dinners with friends. A large kitchen island (in foreground) works for cooking, serving, eating, and a home office. By separating the kitchen from living areas, the island blocks a view of dirty dishes while also creating a focus for gathering with friends.

IN THEIR OWN WORDS

"The simple act of being able to bring my wife a glass of wine using the trolley cabinet is an earth-shattering thing. Being in a wheelchair is uniquely frustrating. But here, I feel this place is on my side. I have a sense of ease....People take ease for granted."—Randy

"I think of the building of this loft as an epic reclamation project. Randy reclaimed his independence. I reclaimed my sanity. And we reclaimed our marriage, one of constant give and take. For that, I am eternally grateful."—Leslie

The loft building in downtown Seattle already had a ramped entrance and elevator, which allowed Randy and Leslie to focus their attention and budget on remodeling the condo interior.

Loft living had great appeal to Randy and Leslie, who were attracted to wide open spaces flooded with light. Rooms just feel larger with high ceilings, and brick walls give a comforting sense of stability: Lofts aren't flimsy. They also wanted an urban setting, for proximity to work and cultural activities. Their search for a new home led to a 1900-era clothing warehouse with an elevator in downtown Seattle, a short distance from shops and restaurants and right on public transportation routes. While they'd lived in lofts before, this was to be their first renovation project, and they sought assistance from Carol Sundstrom of röm architecture.

Their loft was essentially a shoebox configuration, with two windows at one narrow end and an entrance from the common hallway at the other (see the floor plan on the facing page). At 986 sq. ft. the loft was small, so it would be important to minimize space-consuming walls and halls. Both Randy and Leslie liked to cook, so a big kitchen was essential. Carol suggested a linear progression of "rooms" starting at the entrance: from sleeping (far from windows and therefore dark), to a central utility core with bathroom, laundry, and kitchen, and an open living/dining/study area near the windows. The only doors would be at the bathroom and closets.

A multipurpose living and dining room near the kitchen works for casual entertaining and everyday living.

Floor Plan

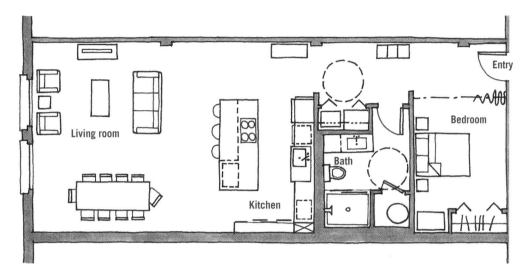

A KITCHEN FOR WALKING, STANDING, OR SITTING

Randy had been a healthy, active person when the couple wed, but when they moved to Seattle he began to need leg braces, a cane, or crutches, depending on distance and degree of difficulty. As construction neared completion Randy was using a wheelchair more and more frequently. Carol knew this home would need to be especially responsive to his changing condition. Working with a local metal fabricator, she custom-designed a system of sturdy handrails that do double-duty as grab-bars and towel-bars in the kitchen and bathroom. At first Randy used the kitchen handrails to steady himself while cooking, then later to pull himself out of his chair to a standing position, and eventually to maneuver his chair about the kitchen by pulling or holding onto the bars.

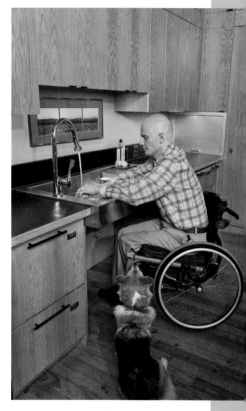

Custom-designed metal handrails at countertop edges support Randy while standing to cook; he also used the rails to pull his wheelchair alongside the island when his hands are too full to operate the wheels. Tall wall cabinets multiply storage capacity within a small condo.

One of Randy's favorite parts of the loft is this custom-designed stainless-steel sink incorporating a drainboard beside a shallow bowl. With a high-arc spout and lever-type faucet placed to the side of the sink, and the surface height and knee clearance sized according to Randy's dimensions, the fixtures are fully accessible.

A curtain takes the place of a door to the bedroom. In the open position (above) the bedroom enlarges to incorporate the hallway for extra maneuvering space. In the closed position, private areas of the loft are out of sight for guests.

DESIGN AS NOURISHMENT

Each stage in the progression of a degenerative condition involves adjusting to new limitations, along with feelings of frustration, anger, and grief at the abilities lost. Asking for assistance was a skill that became easier when Randy realized the neighbors genuinely wanted to help. For both Randy and Leslie, having a home where friends felt welcome and where they each felt comfortable entertaining has been an essential part of the healing process.

Leslie describes their loft as a sanctuary with the kitchen as its soothing hearth. "We eat to live *and* live to eat. We like to feed people, because they feed us. It's all about nourishment." Connecting and contributing to one another is a way of feeling normal. And so the dining table has leaves that can be added to fit a crowd for dinner. Furniture can be easily moved around for parties. Kitchen cabinets convert to wheeled trolleys for easy meal service.

Good accessible design is not simply a matter of removing barriers; it's also about making places where each person's unique skills and interests can find expression. For Randy and Leslie, this translates to a home where they can host events and cooking classes, fund-raisers and dinners. In a diverse neighborhood of artists, designers, and urban pioneers, and in a comforting and comfortable home environment, they have created a life that allows them to nurture friendships and each other, and that, in itself, is what healing is about. ✦

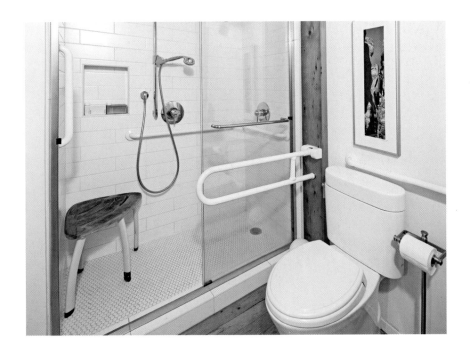

Bathroom grab-bars become a design element when colors and locations are coordinated with other elements in the room. The flip-down grab-bar at the toilet is stored in a vertical position while not in use.

A super-slim bathroom sink from Italy provides knee clearance below and is easy to use in a seated position. Hand towels are within reach, thanks to a custom-designed bar below the vanity countertop. By locating the faucet near an open corner, two people can easily use the same sink from two separate sides. If Leslie is busy showering, Randy can slip into the bathroom to wash his hands, then roll back out of the room.

CHAPTER 11

single-level living

IN 1932 A SAN DIEGO HOUSE RECEIVED ATTENTION for its skillful blending of the Prairie style (integrated into the landscape) with the 1920s Bungalow style (simple detailing), adding elements of the Spanish hacienda (indoor–outdoor living) and western ranch (casual). It was designed by Cliff May, a builder who is credited with over 18,000 tract houses and 1,000 custom homes. By carefully observing how people lived and adding features that enhanced family life and casual entertaining—kitchens with snack bars, floor-to-ceiling glass, living spaces that spilled out onto patios—May's houses came to define the good life in southern California. Many homes featured in this book are reminders that the ranch house continues to be relevant as we enter a new era of designing fully accessible homes.

living big in
a small house

K aren is an inventor, by necessity as much as by
training, and has created a home filled with specially
designed adaptations that suit the three members
of her family. Diagnosed with a condition that gave her fragile
bones as a child and hearing impairment as an adult, Karen is
short of stature and uses a wheelchair. Her teenage daughter
Anita is a paraplegic of average stature. Dave, Karen's husband,
has a unique set of needs in this family, as a person without
physical or sensory limitations. They purchased a mid-century
modern ranch house in the Pacific Northwest because it had
a few features that allowed them to move in without having
to renovate to make it accessible: single-story living, wide
doorways, and a short ramp connecting the garage with the
house, as well as a kitchen and vintage bathrooms with a certain
charm, even if they weren't fully accessible.

An accessible
entrance beside
the garage and
patio opens into
the dining room.

The house is effectively much larger without the walls that formerly enclosed the kitchen. Removing the fireplace that was in the center of the house allowed the kitchen to grow larger. Interior walls stop short of the ceiling to invite generous daylight to flow through to rooms facing north.

Countertops and appliances at various heights define work centers oriented around washing, chopping, mixing, cooking, baking, and cleanup. With simplified cabinetry, the multiple surface heights become an attractive design feature.

Transferring plates from dishwasher to cabinets is easy when everything can be reached from a sitting position.

Educated as an architect, Karen knew the house would eventually need work, but it was her daughter's entry into adolescence—a growth spurt coupled with an active social life—that made the house suddenly seem small. As a homeowner, she'd gotten used to making do, and wanted help in identifying ways to make the house more functional; an addition was out of the question on the small lot. For a fresh perspective on what might be done to gain additional space, Karen called on her colleague Carol Sundstrom, an architect with whom she had collaborated on residential projects.

Carol immediately saw several ways to enlarge the house without changing its footprint. Removing a large fireplace and relocating the adjacent mechanical room would add 100 sq. ft. just where it's most needed, at the center of the house. Moving two interior load-bearing walls would make the main hallway and family room larger, while reducing the master bedroom to a

Floor Plan

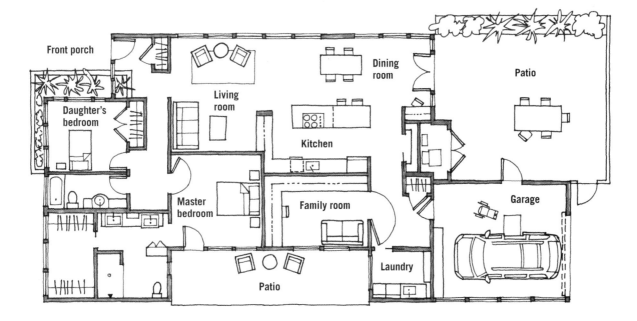

more manageable size. And reconfiguring the kitchen from a U-shaped layout to an L with an island made it more efficient for three cooks.

GETTING THE HEIGHT RIGHT

Karen and her family took on the task of identifying their space needs. Starting with an adjustable-height table and a portable electric cooktop, they set out to discover what kind of kitchen would work for all of them. Their goals were to minimize distances for wheelchair travel, to place appliances and countertops at comfortable heights, to ensure adequate maneuvering space for knees and wheelchairs, and to place storage within reach, whether from a sitting or standing position.

Small appliances come in different sizes and are meant to be used from either above or the side of the item, and so finding the best placement for each object was no small task. They experimented with the height and location of appliances, such as the cooktop, blender, food processor, coffeemaker, and mixer, and with sequences of kitchen activities—chopping vegetables, rolling out dough, pouring hot water into coffee cups, carrying hot dishes from wall oven to countertop, and transferring clean dishes from dishwasher to cabinets. When they were finished with the experiment, each appliance had its own ideal location.

Clearance for knees under both counters and upper cabinet doors was equally important, for Karen's and Anita's wheelchair heights and leg lengths vary greatly. Through a process of trial and error, they identified four countertop heights that would drive the design of their new kitchen: 30 in., 32 in., 33 in., as well as the standard 36 in. In the same way, they experimented with reach ranges and storage requirements for everything from heavy pots and pans to everyday dishes and spices. Eventually they were satisfied they had

UNIVERSAL VS. CUSTOM DESIGN

Many features in the home comply with standards for universal design: A curbless shower and zero-step doors, an open floor plan with plenty of space of move about, and a kitchen designed around various activity centers make this a home that has broad appeal. The sloped garage floor frees up interior space and improves walkway safety (and it also provides positive drainage to an outdoor trench drain). Closets have pocket doors and hardware has easy-grip pulls. For family members and guests, as well as future residents, this is a home that works for everyone.

Appealing as these features are, it is the customized kitchen that makes this house truly outstanding. Cabinets make full use of every inch of kitchen space. Work stations were designed to minimize the need to maneuver about in a wheelchair.

- A clean-up station links the sink, trash/recycle, and dishwasher.

- A put-away station links the dishwasher and cabinets.

- A kitchen island, open on three sides, allows accessible cooking and eating.

- A shallow yet oversize stainless-steel sink maximizes bowl cubic footage.

- An integral drainboard and side-mounted faucets makes washing easy.

- A desk that holds table linens easily morphs into a sideboard.

- Side-opening drawers put contents of corner cabinets within reach and view.

- Pocket doors for low kitchen cabinets slide out of the way when open.

- Cooking appliances have heat-resistant countertops within reach.

- Built-in appliance garages and shelves put small appliances in easy reach.

- Upward-acting pocket doors are out of the way when appliances are in use.

Retractable doors transform cabinets into additional countertop and workspace. A sturdy pull-out shelf holds small appliances, and the television cabinet puts the TV out of sight when not in use. All were designed to clear the lap of a cook using mobility devices.

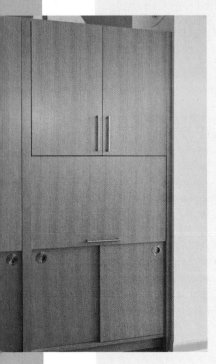

A full-height mirror in a cabinet separating sink and toilet (reflected in the mirror) makes a functional dressing and vanity area for the master bathroom.

This full-height cabinet contains shelves for bathroom storage facing the toilet, and a full-height mirror facing the sink. The owner designed it to be useful for dressing and primping.

the precise knee clearances and reach ranges that would ensure a comfortable kitchen for the whole family.

FAVORITE PLACES: THE BATHROOM

The master bathroom is one of Karen's favorite places in the house. A full-height cabinet creates a private alcove at the toilet area and contains a full-length mirror behind a pair of doors facing the vanity sinks, making it easy to get ready for work or a party. His-and-hers sinks (same design but different heights) share a common countertop for clean lines and easy cleanup. A roll-in shower with a gently-sloping floor was created by removing the curb and adding a trench drain. By placing the shower curtain rod over the sloped surface of the floor, water drips back toward the drain rather than out into the room.

The family room does double-duty for games and studying. An accessible desk sets a visual horizon line that serves to organize storage shelves and cabinets, thereby reducing visual clutter.

GAINING GARAGE SPACE WITHOUT AN ADDITION

The house's former owner had installed a ramp to navigate the 6-in. height difference between garage slab and house floor, leaving the ramp edge as a potential tripping hazard. The slab was repoured as a single continuous sloped surface running from the house to the driveway, making a gently tilted plane for both wheelchair and car ramp. Now there's adequate parking space for Karen's power scooter, with electrical outlets nearby for charging.

Throughout the house, built-in features are tailored to the homeowners' dimensions rather than to standard cabinet modules, and located for ease of use rather than ease of construction. Where the family members' requirements vary, built-in items were sized to work for everyone. The result is a unique aesthetic, a seamless blend of form and function. ✦

accessible ranch house

F or Murray, an aspiring filmmaker who worked at home in the mountainous terrain of British Columbia, an accessible house was essential. But, it seemed prudent to take time to plan rather than rush into renovations after the automobile accident that left him a quadriplegic. Months in rehab had convinced him that he needed a home where he could live independently, one that didn't seem institutional. He felt he had one chance to get the design right.

The first step was to find a suitable home. Murray wanted a forest setting with a view and a sense of privacy, not too far from city conveniences, and on a flat street that was wheelchair-safe. This was a difficult combination to find in North Vancouver, where the land is steep and most homes are split level. When a 1950s ranch house on a quiet cul-de-sac came on the market, halfway up the

A large covered terrace provides protection from rain at the living room's sliding doors. Wide doorways and small low windows give Murray plenty of maneuvering space as well as customized views at eye level.

By removing a central fireplace the rooms are visually linked and the path of travel around the house is widened.

Full-height glass walls define the "livable room nook," while a gas fireplace keeps the space cozy in winter. Low countertops extend the kitchen with an accessible serving counter, and 12-in. toe-kicks and baseboards protect surfaces from damage due to wheelchair equipment.

mountain and near a creek and a ravine, Murray knew this was it, and took the plunge into home ownership.

THE "PRE-DESIGN" PHASE

Murray hired architect Russell Acton, of Acton Ostry Architects, and the two embarked on an extended pre-design phase, sharing ideas and building trust, without the pressure of a construction project. Russell outlined the design process in their first meeting: "You describe the problem, and we'll come up with a solution." For many months Murray lived in the house, paying attention to how light flows through the rooms at various times of day, how the seasons affect daily life, and how the house seemed to fit for various activities. A floor plan of the house, marked up with comments, traces his observations about what needed to change. For example, when friends at a party left dishes on top of the refrigerator or, in attempting to be helpful with cleanup, put things beyond his reach, he knew his kitchen would have lowered surfaces.

He visited accessible buildings such as recreation centers to collect ideas that could be adapted for a home environment. In navigating public places he observed how often he needed to ask people to step aside so he could maneuver in his chair. Take note: His home would have wide halls, doors, and pathways between furnishings. Murray pored over magazines, collecting images. When an opportunity arose to test out his ideas in a vacant house, he took it (see the sidebar on the facing page). Throughout this period Murray made sketches of his ideas. One shows a "livable room nook" where he can sit near the fireplace, looking out to the forest through floor-to-ceiling glass. Another shows long narrow windows at eye level for a continuous view to the outdoors while wheeling through the house. Still another shows a living room configured for plenty of chairs, to put seated visitors at Murray's eye level.

Russell describes Murray as the ideal client—curious, observant, and able to articulate his needs clearly. In time, a strategy for modifying the house came into focus, with a short list of requirements. The house would have an enclosed garage, sized for a van; larger bedrooms and bathrooms, going from three to two; wider halls and doorways, sized so that two people can pass without one

Floor Plan

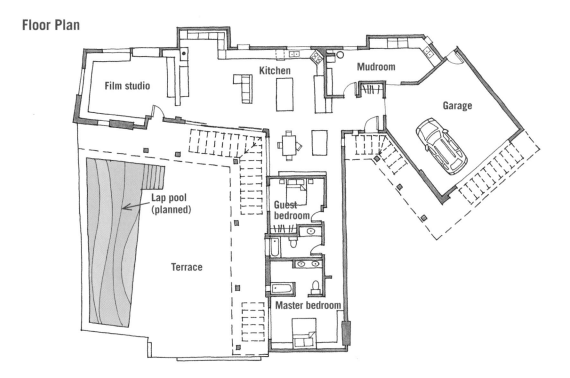

Film studio

Kitchen

Mudroom

Garage

Lap pool
(planned)

Guest
bedroom

Terrace

Master bedroom

having to back up or step aside. In addition, Murray wanted a film studio with 12-ft. ceilings for his equipment. By removing a fireplace the house gained interior space. By expanding a carport, the house gained space for a garage.

"FOR-ME" WINDOWS AND OTHER INNOVATIONS

Murray would arrive at design meetings with inventions from the testing lab. The "for-me" window is one example. He would choose views around the site and frame these with small windows in deep frames, at various heights. For

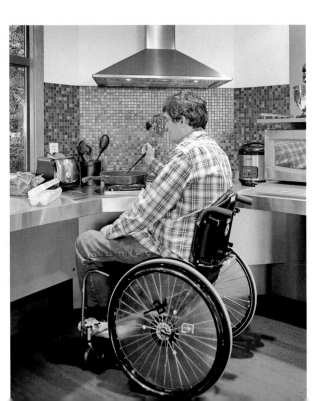

Kitchen counters are stainless steel with a perimeter edge to control spills. A faucet over the cooktop allows the cook to fill pots without dragging heavy containers between sink and stove.

THE TESTING LAB

Murray found a rental house slated for demolition, with an owner willing to let him use it as a testing lab. It was a rare opportunity to track the incremental effects of wear-and-tear, as the wheelchair nicked walls, doors, and cupboards, and to incorporate this information into the design. With help from friends he pulled cabinets off the walls, lowered shelves and countertops, and relocated appliances. He scrutinized every tiny action in his daily routine, from opening doors to turning off lights, and took voluminous notes. When it was time to start planning the renovations, Murray was ready to be a full partner in making design decisions.

"The challenge was to be in this geography but still be able to design a home void of ramps and lifts. I wanted a property that reminded me of Whistler, where I lived pre-injury, and also where I could have my mountain friends over without their feeling they were in a city. It wasn't about making my house comfortable for them, as much as it was about making me feel I hadn't been launched way out of the mountain world by becoming a quadriplegic. And it worked!"

able-bodied visitors who have to stoop down to look outside, these windows offer an insight into Murray's world. A "for-me" window in the garage wall allows help to view inside in case of falls, and also displays vintage cameras—a subtle welcome to the home of a film maker. Other innovations include the following:

■ Secure carpets in transfer areas (sofa, bed) to prevent wheels from slipping.
■ A flat-screen TV in the bathroom to make time there more enjoyable.
■ A raised laundry platform placing appliance contents in view.
■ A faucet over the stove for filling pots.

Murray has adapted to life in a wheelchair, and his house has adapted to his needs. With careful planning, he and Russell have created a house that serves him well. The mark of their successful collaboration is a house with smooth clean lines, spacious rooms, and large windows, an example of universal design that serves everyone well. ✦

"For-me" windows come in various sizes and shapes, but all are designed to focus attention on something unique inside or outside the house, for Murray and his visitors.

Floors are smooth wood for wheelchair travel, but areas of transfer require friction surfaces to keep the chair from slipping. The carpet was set flush with the wood floors, and sized to fit perfectly beneath the sofa, with the same detail used in the bedroom. Radiant heating under cherry wood floors makes for nice even warmth.

Furniture groupings maximize available space for mobility devices and make it easy to expand living and dining areas for larger gatherings

When the garage, driveway, and front door share a common path of travel, site paving and grading are minimized. The result is an entranceway without steps and a driveway that has a level surface to simplify transfer to and from the car.

low-effort living

A steep hilltop might seem an unlikely site for an accessible house. And an outdoor swimming pool might seem a perilous use of land when the homeowner uses a wheelchair or has young children and a big dog. But Elizabeth is someone who chooses to see opportunities where others would see barriers. She wanted a house where the living would be easy, despite the bus crash that left her a paraplegic as a young woman. When she hired architect Jeffrey L. Day of Min | Day to design her house, she challenged him to create a home where she could pretend her disability does not exist.

LOWERING THE FRUSTRATION LEVELS

Elizabeth describes the ways that an accessible home reduces the everyday frustrations that can interfere with enjoyment of living: "Your focus moves away from your disability when activities can be done effortlessly." For example, she finds specialized equipment bothersome, and prefers a manual wheelchair to an electric model—no batteries to charge, no motor to break down, no need to transfer between chairs while batteries are charging. Elizabeth also avoids using a scooter because she finds it impossible to transfer without awkwardness. Despite being situated above Silicon Valley, hers is not a gadget-filled or "smart" house; basic electrical switches and outlets suit her just fine. For Elizabeth, low-tech equipment means low-effort living.

REDEFINING THE RANCH FOR ACCESSIBLE LIVING

At first glance, the house is a straightforward ranch-style house (see p. 114), albeit on a spectacular site. The layout is rational and straightforward, a linear string of spaces with bump-outs at the center, with living areas facing the view (see the site plan on the facing page). Family spaces are connected by open pathways, with storage cabinets that seem to float between floor and ceiling serving to separate travel areas from gathering areas. The bedroom/bathroom wing is to one side of the entrance and can be blocked off by closing a large sliding door.

Custom cabinets define living and circulation areas. In raising the base of this unit above the floor, the design clears wheelchair footrests and allows more daylight to filter into the center of the house. Low shelves lining perimeter walls (at left) provide convenient storage without reducing floor area at the path of travel.

The layout is not only efficient but also functional and playful. Elizabeth appreciates the wide circulation space, and her engineer husband Amardeep appreciates the ways that strategically placed cabinets control clutter. Their two young daughters love having a floor plan with "circles everywhere"—many ways to travel around the house, playing hide and seek with each other and Mojo, their big Bernese Mountain Dog.

DESIGN FOR EFFORTLESS LIVING TAKES WORK

The simple clean lines of the house reflect more than the absence of ornamentation. They are the intended result of a house carefully designed in response to the homeowners' precise requirements for accessibility. Architect and homeowner thought through the sequence of activities in each aspect of daily life. For example, the garage and front door are placed side by side so that one transfer area works for both entrances. Jeff designed a hatch where the garage wall meets the kitchen pantry, so groceries can be easily loaded from the car.

Inside the house, display shelves are strategically located so that there is always a place to put things down within reach, without blocking the path of travel. Cabinets that divide living areas from hallways are raised off the floor to

Site Plan

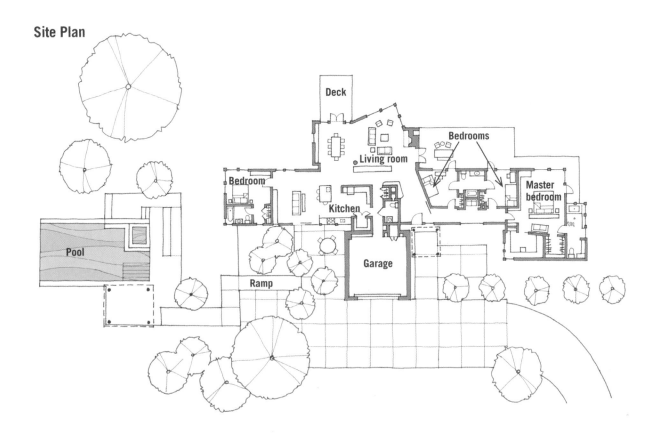

Tile walls and a concrete bench make a low-maintenance shower/bathroom, while nonslip wood flooring improves safety.

provide clearance for wheelchair footrests. A living-room fireplace has a raised hearth, placed to be seen from all living areas, and within reach for Elizabeth. Floor-to-ceiling windows give a view to activities in the yard just outside and to the peaceful hills beyond. In the master bathroom a long concrete bench serves as a tub transfer area and also as a seat for showering. Integrated shower drains under slatted teak-wood floors drew their inspiration from boat design.

FIND IT OR DESIGN IT

Jeff combined off-the-shelf elements with customized installations so that Elizabeth has everything she needs in the house. Appliances include a side-by-side refrigerator/freezer and a side-opening oven. A long stainless-steel countertop purchased from a restaurant supplier is installed at a comfortable height for Elizabeth. With a continuous handrail-plus-towel-bar built into the counter edges, she can easily pull herself between sink and cooktop, or slide a heavy pot of pasta water across the counter. Garage interior walls are finished

A low kitchen island is one of many ways that rooms are divided without the use of walls, to maintain visibility between activity areas. Keeping an eye on what is happening in adjacent rooms is one way to make raising a family a little more effortless.

This stainless-steel counter was purchased from a restaurant equipment supply house. With an edge lip to prevent spills onto the floor or lap, and with an interrupted surface connecting the sink and cooktop, it is easy to push a pot of boiling pasta water safely toward the drain.

with a slatted storage system used in commercial displays, making it easy to keep objects off the floor and within view.

Bathroom sinks are custom-made in concrete, designed with shallow bowls, ample countertop space, and generous knee clearance below. A craftsman was commissioned to fabricate a slinky-type coil to cover the hot pipes below sinks so that Elizabeth's legs are protected from injury. For Elizabeth, freedom from worry allows low-effort living.

STAYING STRONG TO STAY INDEPENDENT

Seven years after the house was completed Elizabeth approached Jeff about designing an outdoor swimming pool so that she could exercise daily. Maintaining upper-body strength is essential for paraplegics, especially for those using manual wheelchairs, and Elizabeth works hard at staying fit. Jeff and partner E.B. Min designed an outdoor swimming pool, re-grading the site to capture flat areas for the pool and surrounding deck. Gently sloped pathways were laid out with switchbacks to form a long ramp, which connects patios at the pool and house in a manner that seems graceful and spacious. Artificial lawn was used around the yard for a flat, dry, mud-free rolling surface that enables Elizabeth to join more fully in outdoor family activities.

Concrete countertops give bathrooms a sturdy functionality but also shallow large-bowl sinks that make light work of cleanup and washing.

An outdoor patio and wide sliding glass doors near the kitchen make it easy to serve meals al fresco. The pass-through counter (above) works equally well for serving and for sending dirty dishes through to the kitchen.

From the kitchen island to open shelving in the family room, from the garage to the laundry, from the patio to the pool, this house abounds with accessibility features. Many are easy to overlook, appearing simply as artful and clever forms, different from what one would expect. This is what Elizabeth wanted from the start: a home where distinctive and interesting elements can make us forget, for a time, that it is also fully accessible. ✦

Curbed concrete ramps weave through terraced yard areas to put all areas of the yard within reach. Artificial turf ensures the ground surface is flat, level, and mud-proof and the yard maintenance-free. The infinity pool beyond is an important part of Elizabeth's exercise regimen for maintaining upper-body strength.

CREATING AN ACCESSIBLE HILLSIDE

Finding adequate flat area for outdoor living on the crest of a steep hill was a design challenge, especially when all living areas needed to be on one level and the site had to be accessible. Jeff carefully contoured the site topography so that everything could be reached by ramps. Areas for indoor and outdoor living are connected by a gently sloped ramp and terraced patios. At the side of the house that juts out over a steeply sloped hillside, an outdoor deck connects all bedrooms with the living room, and a separate deck seems to float over the landscape just off the dining room.

accessible farmhouse and barn

Between the craggy shoreline and rolling farmlands of coastal Maine, architect John Gordon has created an accessible oasis for his daughter. Jessica has been a quadriplegic since college, when a car accident left her paralyzed from mid-chest down and with limited use of arms and hands. The road to recovery was long and challenging, but Jessica has been able to create a richly rewarding life, thanks to a home designed as much to support her abilities as to reduce barriers to mobility.

When Jessica first returned home from rehab after the accident, John designed an addition to the family house, with an accessible bedroom and bathroom and a clear path of travel from the garage. Having a home where she could get around was an important part of her healing, but it wasn't enough; Jessica needed to learn new ways of doing absolutely everything. In a 3-week stay at Atlanta's Shepherd Center, Jessica found new hope and the motivation to live an independent life as a normal young woman in a wheelchair. Upon returning to Maine, she set to work with John designing her dream house on a parcel of land beside the family home.

Jessica's house is designed as a cluster of connected buildings, like the rambling farmhouses of coastal Maine.

Floor Plan

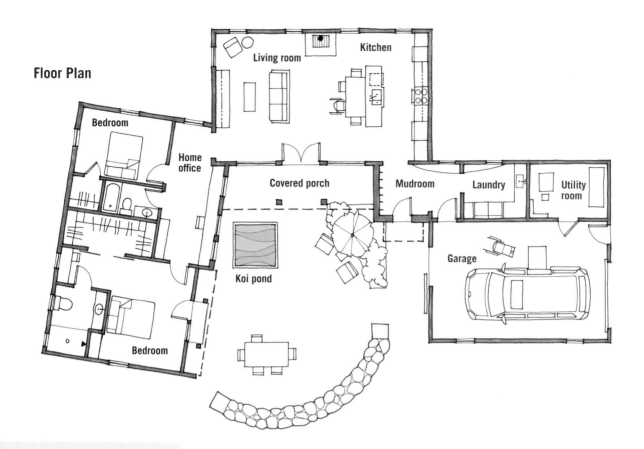

Bedroom

Home office

Bedroom

Living room

Kitchen

Covered porch

Mudroom

Laundry

Utility room

Koi pond

Garage

SIMPLIFYING DAILY TASKS

An important goal in designing for accessibility is to minimize the number of tasks involved in everyday activities so they can be accomplished with ease and safety. Each step that can be streamlined contributes to the independence, and therefore the confidence, of the person with a disability.

Jessica and John have elevated the tasks involved in animal care to an art form, designing a cabinet holding cat litter in the laundry: The cat enters from one side and Jessica scoops out the contents from the other (see the photo on p. 87). Learning to do things differently is the homeowner's job but designing differently is the architect's.

Old Maine farmsteads are a unique architectural style—a string of buildings, barns, and sheds, joined together for travel in inclement weather, built by handy farmers over many generations. Taking cues from local farmsteads, John designed a compound of three buildings linked by a covered porch and hugging an outdoor courtyard. To one side is a red garage, and on the other is a yellow building that houses two bedrooms and bathrooms. A low central building houses common space—a kitchen with open dining and living areas. The entrance is easily found in a tower, reminiscent of both a silo and a lighthouse, with high windows on four sides that beam daylight into a welcoming mudroom below. Forms and materials mirror the Maine countryside—gable roofs, clapboards and wide trim, granite landings, interior plastered walls and painted white casings, wood flooring and cabinetry, all detailed with Shaker-style simplicity.

Paving materials are varied to mark courtyard zones: concrete under the farmer's porch and brick at the courtyard. A blending of covered and open outdoor sitting areas extends courtyard living by offering protected ways to be outdoors.

With the dining table placed close to the kitchen island, meals can be served with a minimum of effort. Low ceilings and focused task-light fixtures bring controlled illumination to daily activities.

Low countertops and appliances within reach combine with drawer-type cabinets and low open shelving to create a kitchen where Jessica can indulge her love of cooking.

A UNIVERSAL DREAM HOUSE

A round paved courtyard draws the visitor into the world that Jessica and John have created. A low fieldstone wall provides extra seating for the gatherings that seem to occur whenever Jessica's at home. With flowering cherry and hydrangea, and buildings on three sides, the courtyard has a cooling shade for much of the day. A sunken fish pond makes a watering hole for dogs Cash and Murphy, Jessica's constant companions. Fronting on the courtyard are four accessible doorways that connect the house with nature and offer a choice in traveling about the home, indoors or out.

The house itself is a skillful blend of universal design and accessibility principles. Wide passageways work as well for Jessica's wheelchair as for her two large and exuberant dogs. Tall windows give seated views to the outdoors and pull daylight through the house. The modern kitchen has a wide island and plenty of storage within reach. In the master bathroom a curbless shower makes for comfortable bathing and dog-washing. Vanity counters in the bathroom and adjacent dressing area provide ample space for personal care. The laundry has a side-by-side washer and dryer, platform-mounted to place doors within easy reach. A wide corridor with a long desk turns what would otherwise be a utilitarian hallway into a dynamic home office for Jessica's graphic design practice. The effect of being in these spaces—colorful and natural, spacious and intimate, and above all, accessible—is somehow both relaxing and energizing.

ACCESSIBLE FEATURES

It's easy to overlook the fact that this house is indeed accessible, but Jessica knows, and that's what counts. Jessica and John visited appliance showrooms to test-drive kitchen equipment and choose those where the dials and handles are easy to use, the drawers and doors not too heavy, and shelves are within comfortable reach-ranges. They designed a pull-out shelf beside the wall oven so Jessica can slide hot pans to one side without spilling the contents into her lap. Kitchen countertops are a convenient 30 in. above the floor, rather than the standard 36 in.

Many of the accessibility features are subtle, like 8-in.-high baseboard and kitchen cabinet toe-kicks that protect surfaces from damage at the level of wheelchair footrests. Maintaining an even body temperature is a constant adjustment for quadriplegics, so John designed the house with in-floor radiant heat. An infrared ceiling panel above Jessica's bed keeps her warm at night. Crank-operated windows can be opened and shut single-handedly. Utility rooms have sliding and pocket doors, and swinging doors at the bedrooms and entranceways have broad landings on either side.

Shower, toilet, and sink areas share a common maneuvering space, making it easier to shift between activity centers.

Corridors can be redundant space unless they're enlivened with new purpose. This hallway between the living area and master bedroom is Jessica's sunny home office and a perch to keep an eye on the entry courtyard.

Across a common driveway sits Jessica's barn; like its counterparts in the region, it's a place that adapts easily between work and play. A wrap-around greenhouse allows Jessica to garden year-round.

ACCESSIBLE BARN

After completing the house, John and Jessica turned their attention to building a real barn across the driveway. Traditional New England barns are exciting spaces, both solid and soaring, used for square dances and for building and growing things, and this is no exception. Jessica's barn is clearly a good-time place. In one corner is the gym equipment where Jessica trains for her team's wheelchair-rugby season. A pool table sits in the center of the space, with room for a crowd or for wheelchairs all around. There's a piano in one corner and a drum set in another. A comfy sofa facing the TV and woodstove seems ready for watching the game, or warming by the fire on a snowy day. The barn is attached to an adjacent service building with a small kitchenette, bathroom, and sauna—all useful adjuncts to activities in the barn. John's pièce de resistance is a wraparound greenhouse where Jessica can work with plants, a fitting metaphor for a home where she has regained her joy in living. ✦

Growing plants from seedlings and bringing color into her house with flowers have been important activities in Jessica's healing after her accident. With wooden walls and concrete floors, the greenhouse is a low-maintenance workshop.

CHAPTER 12

multistory and accessible

IN THIS SECTION WE TIP OUR HATS TO ELISHA Graves Otis, whose invention of elevator safety brakes in 1853 made vertical travel possible for people, not just industrial products. Full accessibility in multistory homes requires either a lift, an elevator, or a very long ramp. Although it represents a major investment in an existing two-story house, adding an elevator essentially doubles usable space by bringing upper floors within reach, making it an attractive alternative to constructing an addition. Elevator placement is a critical decision that affects the vitality of both activity centers and circulation hallways. Whether included in designs for a new house, renovations to an older house, or deferred to the future, planning for an elevator is integral to planning the accessible multistory home.

locating a lift

W|here to place the elevator is a key decision in making a multistory house accessible. When the site is steep or the property is very narrow, it takes real ingenuity to find a location where other rooms are least disrupted. Adding onto an exterior wall is not always an option, especially when the house is close to the property's setback lines and an addition would trigger a zoning variance process. This house on a steep hillside in Oakland, California, presented all these challenges, as well as spectacular views of the San Francisco skyline and Golden Gate Bridge that nobody wanted to block.

A new hallway connects the elevator with the house's main living areas and entrance. Glass doors and a skylight above keep the lift and landing from being dark.

Floor Plan

Upper Floor

Main Floor

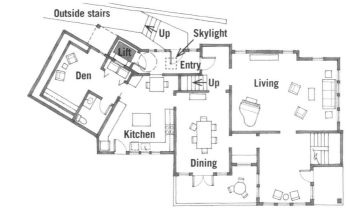

The elevator shaftway wall mirrors the angle of the garage on the upper level. Stone stairs were rebuilt between the street above and a new entrance to the house at the left.

ADAPTING A HILLY SITE

Don was newly diagnosed with ALS (Amyotrophic lateral sclerosis, a degenerative disease of the nerve cells in the brain and spinal cord that control voluntary muscle movement) when he and his wife Diana sought help in making their house accessible. Doctors had advised they move to a flatter neighborhood but, having lived in Oakland for over 30 years, the couple felt it was more important than ever to stay. For Don, moving felt like a decision to accommodate his illness rather than his life.

The house is constructed on five levels, starting with the garage 6 in. above the main bedroom floor—and no space for a 6-ft. ramp to link these two levels. Half a story up is the master bedroom. Living areas—the kitchen, dining, and living rooms—are a full story below the bedrooms. A finished basement holds a family room with exercise equipment and a fireplace. All the upper levels have decks with skyline views. The basement has a densely landscaped patio above a forested hillside that plunges 50 ft. below.

Don and Diana's first priority was to find a way to bring Don into the house, which meant installing a lift to vertically connect the garage, bedroom/bathroom wing, and living levels. Their architect, Catherine Roha, carved out space for the lift along one side of the house, tight against the side setback. Because the three floors to be linked had an overall height of less than 12 ft., the design incorporates a vertical platform lift rather than a residential elevator,

Skylights and doors with glass make the elevator a pleasant and light-filled space. Glass area was limited in the door from the garage, which required fire-rated construction.

FROM A STAIRWAY TO A STAIR LIFT

Connecting sleeping and living areas in the house was a steep stairway with a slight turn in direction just below the upper level. Wall-mounted hardware for the stair lift required a straight run and uniform tread dimensions, so the stairs needed to be straightened and the metal railing extended. After reconfiguring the stair treads and risers, Keith's crew spent ten hours one day welding parts for an extension handrail and installing it.

The main interior stairway was rebuilt and straightened to allow a seated chair-lift to be used during construction of the enclosed lift (behind the glass door beyond).

at some savings in cost. Related work included adding a hallway connecting the main living level to the lift, a new front entrance, and rebuilding the stone stairway from the street to the entrance. Glass doors and a skylight were installed in the lift area to make the small space feel more user-friendly.

DESIGN-BUILD COLLABORATION

Meanwhile, Don's condition was deteriorating rapidly. He went from using a cane to a wheelchair in the five months between when the design started and the installation was complete. In a dynamic collaborative effort, builder Keith Alward coordinated with Cathy on a plan to help Diana move Don about the house while they awaited delivery of the lift. They ordered an inclined stair lift and set about modifying the main stairway to accommodate the lift (see the sidebar above).

Racing through decisions to beat the clock often results in poor choices and higher costs, but with a competent design and construction team, and with clear priorities from the homeowner, the outcome can be successful. This was the case for Diana and Don. By trusting her design/build team, Diana was able to be with her husband at the end of his life. In deciding to stay at home, Don was able to enjoy the places that had always brought him such pleasure, and to know that Diana had a home where she could age well in place. These decisions reduced the stress of living with illness and improved the quality of life for both Diana and Don in their remaining months together. ✦

raised loft

Waking up in the hospital after a motorcycle accident left him paralyzed from mid-chest down, Brian was visited by Bruce Cook, a fellow paraplegic with 20 years' experience of life in a wheelchair. Bruce was a goodwill ambassador with a message of hope, a welcome to the "new normal." Bruce's advice was, "You're going to be making some adaptations but you don't have to do it alone. Learn from those who've gone before you."

At the time of his injury Brian and his wife Donna lived in a multilevel urban loft in Kansas City with an antiquated service elevator dating from 1925, neither of which would work for him any longer. Brian had enjoyed the simplicity, flexibility, and adaptability of an open floor plan. Without walls and doors, which result in separate paths of travel between and within adjoining rooms, there is more usable living area, sunlight flows through the spaces, and

Laminated timber beams above the first floor carry structural loads to the exterior walls and minimize the need for interior walls. Exposed ceiling joists accentuate the implied hallway and accessible path of travel at floor level.

The overall design reflects Brian's accessibility goals—multilevel living, maximum usable outdoor space, and a fully accessible interior modeled on loft living. Natural exterior materials respond to the character of neighboring bungalows and also the owner's wish for a home that transcends the traditional.

activities can be expanded or contracted simply by shuffling furniture around. Now that he was disabled, Brian knew that eliminating walls would remove many of the obstacles in navigating around the home.

LOFT INTERIOR, TRADITIONAL EXTERIOR

Their house hunt led to a vacant pie-shaped lot in a Missouri neighborhood. A state-designated historic district filled with two-story bungalows, the site seemed, at first glance, an impossible place for an accessible home. But the Historic Landmarks Commission was encouraging: Regulations discouraged imitating historic styles but they welcomed contemporary interpretations of historic exterior details, and had no stipulations about the interior layout.

Countertops at two levels and wall ovens reachable from a seated position make this a kitchen where cooks with different physical requirements and abilities can easily work together.

Floor Plan

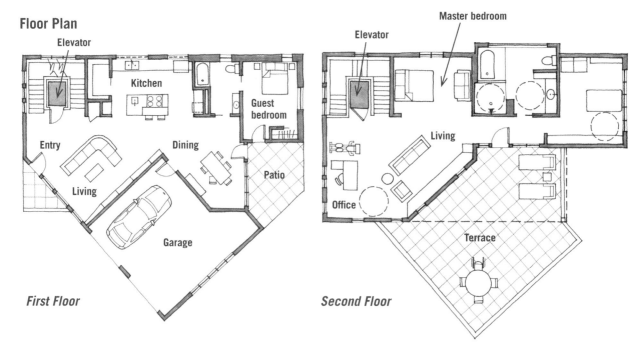

Elevator
Kitchen
Guest bedroom
Entry
Dining
Living
Patio
Garage

First Floor

Master bedroom
Elevator
Living
Office
Terrace

Second Floor

Encouraged by the good news, Brian and Donna retained the design-build firm G3 Collaborative and, working with architect Lon Booher, set about creating an accessible two-story loft, disguised as a contemporary residence in a historic district.

A LAYOUT TO MIRROR THE SEQUENCE OF DAILY LIVING

A disabling injury requires a person to carefully relearn all the mechanics of daily living. For the able-bodied, activities such as bathing, dressing, and brushing teeth function as if on automatic. In rehab, Brian had plenty of time to scrutinize how he did things and by the time they started to design the house, he was an expert regarding his new limitations and abilities.

The dressing room is a paragon of ergonomic functionality. The same space near the master suite doubles as a laundry, dressing room, walk-in closet, and stretching exercise room. A futon platform with storage below is a good place to fold clothing and store linens.

As he describes the house, it was "designed as an idealized diagram of ergonomic function." For example, upon waking Brian travels the short distance from the master bedroom to his futon, located in the center of a dressing/laundry room, then transfers to his shower chair, showers in a roll-in shower, and returns to the futon, on which he dresses and begins his day. Clothing is stored in shelves, drawers, and closets lining the room. The front-loading washer and dryer is in the dressing room, enabling him to do the laundry while showering. The futon is located on a platform, level with wheelchair seating and with storage for seasonal clothing beneath. This arrangement serves multiple functions: laundry folding table, stretching/exercise mat, and dressing deck.

Interior walls define the stairwell and bathroom on the second floor. Note the metal lattice headboard, which makes it easier for a paraplegic to adjust his position in bed as well as to transfer to and from his chair.

ACTIVE AND PASSIVE LIVING

Before his injury Brian worked as a landscape architect, so conversations about place and space are familiar territory. He believes successful house planning has two parts. Passive components involve staying in one place and include eating, reading, writing, watching television, surfing the Internet, sitting, and sleeping. Active components involve moving about and using tools and equipment: bathing, showering, brushing teeth, toileting, dressing, doing laundry, cooking, housekeeping, yard work, and so on.

Passive features in the house comply with good standards of universal design: zero-step entries, wide doors and hallways, a level path of travel around rooms and between indoors and out, knee clearance at countertops and

IN THEIR OWN WORDS

"In many regards we designed around my injury as if building to an idealized diagram. From getting out of bed, to getting into my roll-in shower, to getting dressed and out of the house, we thought things through, we let form follow function, and effectively removed common barriers/obstacles such that I live a life of relative ease and comfort."

A large roof deck over the garage maximizes outdoor space on the site. The deck is easily accessed from the second floor master bedroom area.

toe clearance at cabinets. In addition, Brian planned the placement of built-in furnishings and the layout of rooms to make it easy for him to get around with a minimum of effort. For example, the master bedroom has a small kitchenette with a wetbar, electric kettle, and built-in refrigerator, so once he's upstairs he doesn't have to return to the kitchen for refreshments. There is a large roof deck over the garage, beside the bedroom and level with indoor flooring, so Brian can easily roll outside.

Active daily living is more complex, and involves selecting easily used fixtures and appliances, as well as designing built-in furnishings to

accommodate Brian's unique needs and physique, while being usable by his wife as well. The laundry has a front-loading washer and dryer, and space nearby for an ironing board and drying rack. Kitchen appliances include a drawer-style dishwasher, an insta-hot water dispenser at the sink, and an in-wall convection oven tailored to Brian's reach. Built-in items include drawer-style base cabinets, a pull-out shelf beside the wall oven, and pull-down shelves at the wall cabinets. Ceilings in the bathroom were structured to allow future installation of a mobility lift.

A wide tub deck creates a comfortable transfer platform beside a curbless shower. Shelves and grab-bars within easy reach make bathing both private and safe.

OPTIONS FOR PLACES-TO-BE

Home elevators are no longer an extravagance but, rather, a rational response to both the homeowner's physical limitations as well as the site constraints that result in multistory houses on small urban lots. When vertical travel is accessible, the distance between rooms is shortened. Brian's elevator more than doubles usable outdoor space on the property, allowing him to reach the garage roof deck, thereby allowing the yard to be bigger as the house's overall footprint is smaller. For Brian and Donna, installing an elevator in a house built from scratch was a no-brainer; the elevator restores Brian's independence and makes available the full range of multistoried architectural options for their new house. ✦

urban infill

A ccessible homes need not be limited to rambling suburban lots; they can also be multistory structures on urban lots, marketable to a broad cross-section of the population. In designing a house for his young family, architect Emory Baldwin's intention was "to demonstrate that universal design can blend in with the rest of the neighborhood and not appear unusual in any way." The house stands three stories tall, including a finished basement designed as an in-law apartment and accessible at the backyard via a public alley. Wide doors and hallways throughout the house offer enhanced maneuverability, as do open-plan living rooms and zero-step entrances.

The second floor can be expanded over the dining area for added extra space without an addition.

From the street, the house blends in with the neighborhood, hiding the fact that this is an unusual building: universally designed and accessible.

LIFECYCLE PLANNING

Emory wanted a house that would serve the family through life's inevitable transitions and minimize the chances that they would have to move. The basement is important swing-space, with a compact kitchen, an accessible bathroom, and a bedroom with sliding doors that link it gracefully with the adjacent living room. This apartment is equally suitable for long-term guests or a live-in nanny, a college graduate returning home to get her bearings, or aging parents. In a poor economy, it can be rented out to provide extra income. Over the long term, it's a small apartment for an elderly homeowner with grown children and grandchildren living above.

Wide openings between rooms on the main floor and an island kitchen make the house visually and physically accessible. Parents can keep an eye on young children at play, who are comforted knowing an adult is nearby.

THE LURE OF AGING AT HOME

Let's face it: Moving from one home to another is disruptive and exhausting, and becomes even more difficult when a family's roots in the neighborhood run deep. Most Americans expect to stay in their home as they age, although when they do move, it's usually either to downsize or reduce maintenance chores by moving into a newer home. The idea of living on one level drives 60% of decisions to move, a percentage that increases with age.

We've seen property values plummet in a weak economy, and many homeowners stay put in a house that no longer meets their needs because they can't afford to sell. With a house that allows his family options for living at home, architect Emory Baldwin hopes to reduce the stress that can accompany an illness, injury, or the normal course of aging. Whether the basement apartment is used for a caregiver or for tenant income, Emory has designed a cushion to buffer the family against changes in health or the economic climate.

The bathroom vanity cabinet can be easily rolled aside so that sinks can be used from a seated position.

Flexibility is designed into the main home as well. A two-story dining room allows a bedroom to be added in the future simply by filling in the floor above it. Bathroom vanities have removable base cabinets on lockable caster wheels that allow them to be rolled aside for a resident who may need to use the sink from a seated position. Beside the kitchen is a multipurpose space that is equally appropriate for overseeing young children at play, giving teens a place to gather in groups, or as a crafts studio or homework center. As the children grow up and move away, the space will morph seamlessly for untold other purposes.

The main stairway wraps around a shaft structured for future installation of an elevator. Designed as short runs of five to six steps each, with sturdy handrails and step-lights, the stair provides for safe vertical travel by foot.

The central stair wraps around a tower of stacked rooms, sized and wired for future elevator installation. On the lowest level, the tower holds storage. On the topmost level, the tower holds a reading nook between bedroom areas.

Floor Plan

Upper Floor

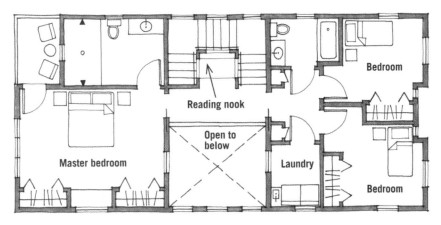

Reading nook

Open to below

Master bedroom

Laundry

Bedroom

Bedroom

Main Floor

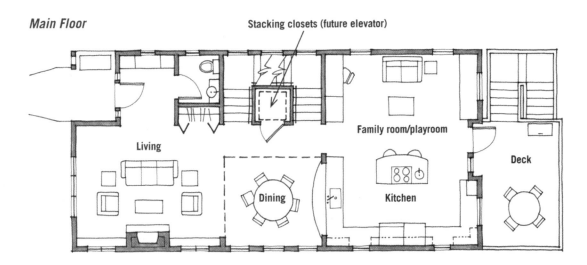

Stacking closets (future elevator)

Living

Family room/playroom

Dining

Kitchen

Deck

Outdoor living spaces coordinate with a future elevator to put nature within reach at each level. A gently sloping path to the left links the basement apartment with a public alley running alongside the back yard.

Currently fitted with three stacking closets, the shaft is framed, sized, and wired for a future elevator cab, but for now, the top closet is a playful "crow's nest" with lookout windows and a built-in bookshelf. By grouping the stairs and elevator together, both share landings at each floor, which eliminates any disadvantage to one means of travel over another.

UNIVERSAL DESIGN

In planning a home where the residents could age in place, Emory incorporated principles of universal design. Outside pathways slope gently up to the house from the street at the first floor and from the alley at the basement floor, at a pitch of less than 1 in 20. All passageways, interior and exterior, have level thresholds and opening widths of 3 ft. or more. Both door hardware and plumbing fixtures have lever handles, and cabinet pulls can be opened easily without grasping. Showers are curbless and shower controls are offset, so a caregiver can turn on the water without getting wet. Thermostats and other controls have large lettering and are located 48 in. to 54 in. above the floor, within viewing range from a seated or standing position.

Universal design features function equally well for people of all ages and abilities. For example, wide doorways are equally convenient for moving furniture or maneuvering in a wheelchair. Wheeled mobility devices, from baby strollers to chairs and scooters all navigate the house's zero-step entrances in safety and comfort. Radiant in-floor heating creates steady indoor temperature levels and eliminates the need for radiators, ducts, or vents, while also expanding options for locating furniture. With loop pulls on kitchen and bathroom cabinets, a person with limited dexterity or with their hands full can manipulate doors and drawers. Decorative rocker-style light switches are attractive and can be operated with the wrist or elbow.

With flexible spaces and universally designed details, this house will be a good fit over the family's lifecycle, and at the same time, will appeal to a wide range of potential buyers, should circumstances lead them to sell. By moving furniture rather than walls, the homeowners can adapt the house to changing lifestyles without a costly renovation. In setting the goal of designing a house that's subtly accessible yet blends into the Seattle streetscape, Emory has created a truly unusual home—and at the same time, one that seems entirely familiar. ✦

High ceilings make small bedrooms seem larger.

Low cabinets flanking the fireplace provide accessible storage and displays; they also allow generous windows to be placed above, for increased natural lighting.

house for clean living

N ew cars literally make Barb sick. Off-gassing from new materials triggers her multiple chemical sensitivities (MCS), resulting in migraines and mild cognitive impairment. Barb also has an inherited condition called multiple epiphyseal dysplasia, which caused her joints to grow incorrectly and limited her adult height to under 5 ft. Repeated hip surgeries have helped her walk with less pain, and with occasional use of a wheelchair. She also has a scooter and van-lift for long journeys.

Barb and her husband Hans decided that a new home offered the best chance for chemical-free and accessible living. Under a doctor's guidance, they embarked on a program to test construction products and choose the least disabling materials (see the sidebar on the facing page). At the same time, they contracted with McMonigal Architects to begin planning. Rosemary McMonigal recalls the couple's wish list and its implications for the design:

- Promote togetherness, so that everyone uses the same path of travel.
- Let the sun shine in. Maximize daylight and natural ventilation.
- Plan for accessibility and adaptability as Barb's needs change.
- Accommodate variations in user height, for both tall Hans and short Barb.
- Choose materials and finishes that do not trigger reactive symptoms.

A farmer's porch connects the house (right) with the garage (left).

Floor Plan

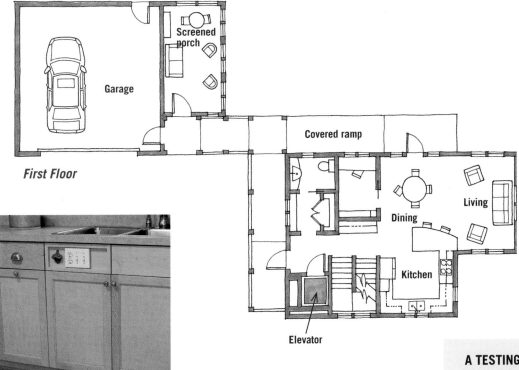

First Floor

A pull-out step lets Barb get to the sink and dishwasher, while outlets on the face of the cabinet put electrical controls within reach.

Rosemary chose an L-shaped layout to give all major living spaces windows on three sides. She centralized circulation using a scissor-type stair with an adjacent elevator, and she separated the garage and house to keep automobile fumes away from living spaces, linking both areas using a zero-step covered pathway for all-weather protection.

The environmental heroes in this house's story are the building systems (heating, ventilation, cooling, plumbing, and electrical systems) and materials selection (nontoxic finishes, no-biocide paints in neutral colors, and galvanized metal). An air-to-air heat exchanger warms the exhaust air, tempering the make-up air entering the system from outdoors. With six control zones, the number of air changes per hour can be fine-tuned for days of high allergen content or for extra indoor humidity generated through cooking, showering, and other activities. There is a whole-house water filtration system to remove chlorine and other chemicals from showers, laundry, and cooking. A reverse-osmosis system cleans drinking water using pressurization to remove undesirable chemicals and biological contaminants. House heating is electric, as natural gas was one of the strongest environmental triggers of Barb's MCS. Appliances and the domestic hot water heater are electrical as well.

At only 880 sq. ft. per floor, the house plan is compact, but well-designed details combine with a clean environment to meet all the homeowners' needs. ✦

A TESTING PROTOCOL FOR CHEMICAL SENSITIVITIES

Many common building materials contain and release toxic chemicals that cause severe allergic reactions for those with MCS. Barb and her medical team developed a kinesiology protocol for identifying and avoiding these in construction. Paints, adhesives, laminates, and finishes were applied to materials to be used in the work, such as wood or drywall, and then either left in the open or placed in a closed jar for a period of time. Barb would open the jar and breathe the concentrated vapors, and her chiropractor would manually test her arm strength, both before and after exposure. Materials that caused any reduction in muscle strength were not used in construction. Two years of research paid off: The new home environment does not trigger Barb's chemical sensitivities.

CHAPTER 13

vacation homes

A VACATION GETAWAY—BESIDE A LAKE, IN THE mountains, or near the shore—is a dream that can seem especially remote when one has a disability. The same principles that apply to making year-round homes accessible and to creating places for outdoor living are especially applicable in the design of vacation homes. We want our getaway places to connect us with nature, both visually and functionally, and to provide opportunities for relaxation both alone and with others. Lay out the vacation home for effortless living and create details for low maintenance so that it is truly a place where the living is easy.

The porch extends beyond the building wall to capture lake views. Sitting surrounded by gardens and within sight of the driveway, Janis can keep tabs on the natural world as well as arriving guests.

lakeside cottage

Based on her daughter's enthusiastic vision of hosting family summer vacations by the lake, Janis bought the waterfront property sight unseen. The house was tiny, one bedroom and 616 sq. ft.—what the local folks called a "camper," and just the kind of country retreat Janis had long been searching for as an escape from the pace of life in Manhattan. The photos her daughter emailed were not impressive, but waterfront property was hard to come by, and as an interior designer Janis could see its potential. The cottage would have to be accessible, for Janis had been using mobility devices since a car accident 30 years ago. She had designed her apartment in Manhattan with built-ins everywhere to maximize usable space and knew she could do the same here.

Although Janis's design career was not focused on accessibility, she was an ardent advocate for universal design in all her work (see the sidebar on p. 158). Most of her clients are initially wary of making their homes accessible, thinking, "It won't happen to me." Janis asks them their parents' ages, and whether they ever visit. Then the conversation turns to price: Most clients fear that universal design will cost them more. Janis has to convince each one

With a large island used for meal preparation, dining, and computer work, and a wide opening to the new family room, the kitchen feels gracious as well as functional.

that a bathroom renovation doesn't cost any more because a door is 4 in. wider. In the end, Janis says simply, "Aesthetically, it just looks better," and the homeowner finally buys the idea.

THE BASICS: AN ACCESSIBLE ENTRANCE, KITCHEN, AND BATHROOM

The first step in transforming the cottage was to make it wheelchair-accessible. An interior stairway leading to the basement is unnecessary when the homeowner uses a wheelchair—and the basement could be reached from outdoors. Simply removing the stair greatly enlarged living space. Floors were made level, without steps or thresholds. Janis designed a new kitchen, with

Kitchen appliances were selected and installed according to intelligent and commonsense principles. The refrigerator door clears the knees of a user in a wheelchair, while the freezer contents are within reach. A single-drawer pull-out dishwasher is easily loaded in a single motion. The wall oven is placed to align with the countertop, making a place to put hot pots easily

Floor Plan

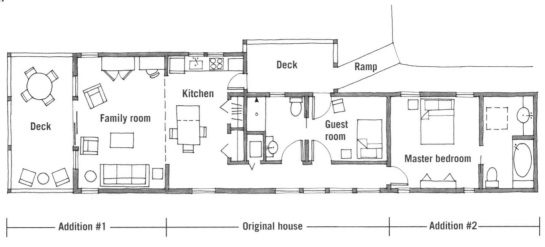

├─── Addition #1 ───┤ ├─── Original house ───┤ ├─── Addition #2 ───┤

29-in. countertops and a work island that doubles as a dining room table. Base cabinets have drawers instead of doors, so the contents are easier to reach, and food is stored in a well-equipped pantry. There are no wall cabinets, leaving room for a large window over the sink.

Initial renovations enlarged the only bathroom by borrowing space from the adjacent bedroom. There is a roll-in shower with a seat and hand-spray, and a tilted mirror over the sink. The toilet seat is a little higher than standard. Doors to the hall and bedroom occupy side-by-side corner walls, an efficient use of space since only one door would make the corner otherwise unusable. Janis added an exit doorway at the bedroom in case of emergencies.

THE LUXURIES: A LARGER LIVING SPACE

Initial renovations also added a family room with a deck overlooking the lake. With cathedral ceilings, sliding glass doors to the new deck, and a zero-step entrance, the house makes outdoor living easy and accessible. Daylight and natural ventilation bathe the space, and ripples on the lake cause dappled light to play across interior surfaces. Generous hallways and comfortably sized rooms make the cottage seem larger than it actually is, and give Janis's young grandson freedom to construct block-cities that stretch across the floor.

PLANNING FOR THE LONG TERM: MAKING ROOM FOR ASSISTANCE

After Janis broke her leg in a fall, she decided it was time for a personal care attendant, and set out to create a second bedroom. By designing an addition to house a modest master bedroom/bathroom, and borrowing 4 ft. from the cottage's only bedroom for a hallway, she was able to achieve her goals. The new bathroom has a tub and a sink set into a narrow counter. A skylight in the bathroom and high windows in the hallway bring in natural light while maintaining privacy.

The original cottage bathroom was enlarged and updated to create this guest bathroom. A tilted mirror, a wall-hung sink, a curb-less shower, adjustable hand-spray, and grab-bars combine for a functional and accessible bathroom.

Accessible homes can feel spacious without being large. High ceilings and full-height glass connect the new family room with both the sky and the lake outside. Parsons tables replace desks for accessible workspace.

UNIVERSAL DESIGN EQUALS INTELLIGENT DESIGN

It just makes sense to design spaces that everyone can use. Wide doorways and zero-step entrances are easy to maneuver, whether in a wheelchair, with bags of groceries, or pushing a child stroller. When kitchen and bathroom storage compartments are within easy reach, young children and the elderly can use them. Lever-type faucet handles and door hardware work as well for arthritic fingers as for the cook whose hands are dredged with batter. Generous, curbless showers limit falls and make floors easier to clean. Janis approaches every design project with four basic principles in mind:

1. Avoid small doorways.
2. Avoid low toilet seats.
3. Avoid wall cabinets in kitchens.
4. Design large pantry spaces.

Controls at the front of the cooktop can be easily used without reaching across hot surfaces and boiling pots. Tile countertop surfaces make it safe to put hot cookware down without scorching the finishes.

INTELLIGENT DESIGN LEADS TO AESTHETIC CHOICES

In every detail the cottage reflects Janis's personal philosophy of intelligent design. Spaces have a serene quality, the result of careful planning for storage. Janis chooses high ceilings and large windows for the health benefits of sunlight and natural ventilation. She also specifies natural finishes and materials to reduce the chances of introducing toxic materials into the living space. Floors are wood, countertops are stone, and cottons, wools, and linens are used for upholstery, shower curtains, and lampshades.

Doors and hardware selections are based on simple physics: Most people find it easier to use push-pull hardware than to turn a knob, and so closet doors are bi-folds. Janis uses pocket doors where they can be left open and where maneuvering space is at a premium, especially in bathrooms beside bedrooms. She designs her kitchen base cabinets with drawers to avoid the discomfort of kneeling down to rummage about in deep shelves. For furnishings she uses Parson's tables instead of desks and bureaus because she finds it easier to sit with legs underneath and to set the surface at a comfortable height.

Years of living and designing in Manhattan have inspired Janis to find ways to use small spaces efficiently, a skill that would be needed for renovations at the cottage. In the master bedroom Janis designed an armoire with low closet poles and a top surface where she can place items for display. A utility closet, food pantry, and laundry are also customized to each unique function.

Art has always been an important part of Janis's design practice—displaying it and seeking out artisans who can fabricate special pieces. After the functional decisions have been made, what distinguishes one home from another comes down to personal taste. Janis believes the home is a private haven, which should have things that give pleasure and comfort, pleasing textures and colors that make the person living there feel good. Janis's tasteful, artistic, and intelligent vision is apparent throughout the cottage, all 1,000 sq. ft. of it. For Janis, artistry and intelligence are the keys to successful accessible design. ✦

Customizing the closet to both the user's dimensions and their clothing needs economizes on space. In this custom-designed armoire, the top surface is within reach for display and a television set.

Simple touches help create an accessible master bathroom. Pairing a narrow countertop with a cantilevered sink results in a comfortably sized basin and a shelf where nothing is beyond reach. Setting shelves near the tub puts towels nearby for bathing. Skylights rather than windows bring in natural light while maintaining visual privacy.

universal design in a second home

U niversal design held the key to solving space problems at this cabin nestled in the Berkshire Mountains of Massachusetts. Joe and Kristin loved having a second home where their three children could enjoy the outdoors year-round, but the split-level cabin was definitely quirky. The main entrance was up a half-story and led directly into a tiny kitchen—an awkward point of arrival with luggage and ski gear. Bedrooms were on two levels, with three at the basement and one at the second floor. At mid-level there was a tiny living/ dining room and a small deck. The couple turned to designer Josh Safdie at the Institute for Human Centered Design (IHCD) for help in making the cabin large enough for overnight guests and more user-friendly.

Open shelves flanking the fireplace are raised off the floor, leaving room to clear a footrest on wheeled mobility devices while also making it easy to clean below. A single sliding door at the games shelf blends easy access with concealed clutter. The central stair was rebuilt with shallower steps and sturdy handrails.

Simple shed forms and natural wood siding are typical of New England mountain cabins.

IN THEIR OWN WORDS

"Our primary home is in Brooklyn, New York, and it's a narrow three-story building all chopped up into rooms. We love the open feel of our mountain cabin! What we wanted when we started the project was to get the most benefit from the least amount of living space, and this is what Josh's design gives us. It's a great house for entertaining our friends who flock here on weekends, and wonderful when it's just our own family. We're even ready if, or when, someone breaks a leg skiing!"

Josh introduced Joe and Kristin to the idea of universal design, as an environment that works for everyone. The basic concept would be an open floor plan with living spaces on one level, ramped if necessary but without stairs—an idea that made sense and felt safe with three active kids. When he explained that universal design would not increase construction costs, Joe and Kristin were sold. They had visions of sharing the cabin with their own aging parents and, later, their children's children. With universal design, they could live in the cabin well into their later years.

FROM PINWHEEL TO TELESCOPE

The original cabin was oriented vertically, with rooms pinwheeling around a central stairway on three levels. The new cabin is oriented horizontally; living areas are telescoped together, one flowing into the next, beside a stacked cluster of bedrooms. Josh redesigned the central stairs using a shallower pitch, adding handrails and lighting. There are two parts to the addition: an extended family room and kitchen at one end, and a gently ramped entrance with

A short interior ramp connects the mudroom with the main living space, giving views into an upper-level balcony and the basement home office (at right) and the kitchen beyond.

mudroom and powder room at the other (see the floor plan on the facing page). The new bathroom was plumbed so that a shower can be easily added, and the layout anticipates adding a future bedroom for single-level living.

Good things happen with the new layout. First, the entrance level is safely near ground level. Second, the linear layout makes a straight path of travel with clear sightlines between activity centers. Third, the plan avoids interior walls between living areas, creating one long and narrow multiuse space—cozy for family time and expansive for entertaining guests.

The design controls indoor climate naturally. Radiant heat in the family room makes the floor warm to the touch and avoids space-taking heating appliances. All new rooms have windows on opposite walls, inviting natural light and ventilation throughout the day. A new fireplace beside the stairs wafts heat up into the bedroom wing on cool nights, while skylights pull cooler air from lower windows and out through the bedroom wing on warm nights.

UNIVERSAL DESIGN EQUALS COMMON SENSE

Accessibility features are everywhere, but so seamlessly integrated into the design that the cabin appears simply to reflect the preferences of five people. This home has an informal friendly feel, inside and out, designed in response to four-season New England lifestyles, from camping and skiing to hiking and cookouts. The good news is that those same features that meet the needs of an active family also work for a wide range of users with differing abilities:

Low open cabinets make it easy for children to help with kitchen chores, and put storage within reach for a person in a wheelchair.

When the dining table is near the kitchen, an island doubles as workspace and serving counter.

The first-floor powder room is designed to be visitable. The wall to the right was plumbed for future addition of an accessible shower, taking the place of an outdoor storage shed.

- A generous mudroom with sliding closet doors allows several people to bundle up at once; it is also large enough for wheelchair turns.
- Wide aisles around the kitchen island make space for lugging groceries and for cooking together—and also for wheeled mobility devices.
- Low windows frame views of chipmunks scampering along the forest floor, from a seated position.
- A screened porch level with indoor living areas allows safe walking with a tray of food and drinks, or safe travel with mobility aids.
- The kitchen has an open counter with a roll-out serving trolley, or stool space for a seated cook.
- Cabinets beside the fireplace have low shelves, good for small children hunting for a board game—and also for a seated user retrieving a book.

Planning for our needs at the moment is short-term thinking. Our homes will last for decades. It makes sense, then, to design for the long term. The truth is that all of our needs and abilities change over time, both in the normal sequence of growing up, as well as in response to life's challenges. With an accessible home, the house gracefully adapts to these realities. ✦

Floor Plan

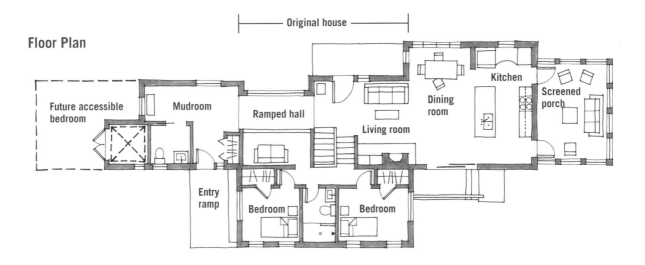

Original house

Future accessible bedroom

Mudroom

Ramped hall

Living room

Dining room

Kitchen

Screened porch

Entry ramp

Bedroom

Bedroom

CHAPTER 14

child-centered homes

IN THIS SECTION WE LOOK AT ACCESSIBLE HOMES for families with children, where either a child or a parent—and in some cases, both—has a disability. Designing for a family means anticipating the changing needs of growing children and their exhausted parents. Sibling dynamics also play a role, requiring both equitable private spaces and adequate shared facilities, which may be a little larger than usual to allow for both active recreation and assisted care.

When a parent has the disability, children step up to help with household chores, and so dimensional clearances and reach ranges may need to be tailored to children rather than adult users. Designing for a family means considering the needs not only of individuals but also of their patterns of interaction; designing for disability raises the bar for creating appropriate environments because the inhabitants are less able to make adaptations on their own.

bathroom for mother and son

A little pampering goes a long way when coping with life's challenges. For Gayle and her family, caring for an autistic son with cerebral palsy is a full-time job. Gayle and Davis have a special relationship. At age 12, Davis is largely nonverbal and uses a wheelchair; he relies on his mother's assistance for all the activities of daily living. Their bathroom on the first floor of a 200-year-old house is a place where Gayle and Davis spend a lot of time.

Concealed within the lines of trim, the mobility lift sits unobtrusively against the wall between toilet and tub. A lightweight motor and sling can be stored elsewhere and hooked up to the lift when needed. A recessed storage cabinet near the toilet puts diapering and clean-up supplies within easy reach.

By removing interior walls and reconfiguring the dressing room, Stephanie was able to create a larger bathroom. The toilet gains privacy from its placement behind the closet wall. Light marble and dark wood finishes offer both brightness and improved visibility in a room with few windows.

Gayle approached Stephanie Gilboy about designing the bathroom with her short wish list: a shower large enough for two people and a rolling bathing chair, a lift to get Davis in and out of the tub, a Washlet toilet with cleaning supplies close by, and two sinks: one accessible and child-sized, the other with a large mirror and vanity. She also wanted in-floor heating and a heated towel bar, as Davis's slight frame chills easily. And somewhere near the toilet she wanted a changing table. The room's footprint could not change.

Beyond these functional requirements was a more subjective list. Gayle wanted her bathroom to be luxurious as well as convenient. Her husband Mark was in full agreement. He felt his wife had worked so hard and long with Davis that she deserved the Ritz: crystal chandeliers, marble floors, rich wood hues, the whole nine yards. And because of the additional challenge that Gayle is visually impaired, the bathroom had to be a place where creative design made it easier for her to see.

GETTING IN AND OUT OF THE TUB

Davis is happiest submerged in a few inches of water, without his chair, free and independent for a part of each day. Splashing around, his spastic muscles relax and he becomes calm and happy. Gayle wanted a deep soaking tub for herself too.

Getting Davis in and out of the tub was becoming more of a challenge as Davis grew older, and Gayle couldn't take the chance of dropping a wet, slippery body. Stephanie loves a design challenge and set herself the goal of finding a transport system that was both beautiful and useful. There were several cumbersome portable lifts on the market that would need space for parking when not in use. Built-in lifts comprising a large framework with attached motors and slings seemed intrusive. Most devices seemed institutional, and Gayle didn't want her guests to feel uncomfortable using the bathroom. Stephanie's research paid off; she found a lift system that fits tight to the wall, with a removable sling and lightweight motor for easy transport to other parts of the house (a future pool, for example).

DESIGN FOR CLEANUP

A changing table for an adolescent occupies a lot of space. Stephanie sketched fold-up and drop-down devices, then plunged into the Internet to find something that might work. She found a tri-fold upholstered panel system designed for lounging over a whirlpool tub—an undermount system that seemed adaptable for this bathroom. Stephanie detailed a lip around the tub's marble platform, then secured the panel's flanges over the plywood substrate below it, and sealed the edges for a water-tight installation.

DESIGN FOR VISUAL LIMITATIONS

Color contrast makes it easier to see, so bathroom finishes use dark walnut paneling and variegated marble. Plentiful light is also important, so Stephanie installed frosted-glass doors at the windowed closet and fixed glass in the shower's interior wall. White plumbing fixtures brighten the room and reflect light, complemented by sparkling polished nickel fittings. A brick wall, the original building exterior, clearly marks the bedroom entrance. The finished bathroom is a place that makes life easier, and happier, for all members of the family. ✦

Bathroom Floor Plan

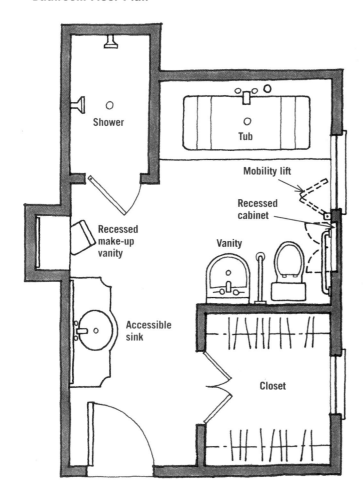

Shower

Tub

Mobility lift

Recessed cabinet

Recessed make-up vanity

Vanity

Accessible sink

Closet

Luxurious materials contrasted with rough brick make a bathroom that appeals to the senses.

indoor-outdoor living

Joanne and Peter had the advantage of having adapted one house before they tackled this one, and they knew from experience the drawbacks to renovating an older building. Their first home was a gambrel-roofed Colonial with modest-sized rooms linked by hallways. They took on the project of making it accessible when daughter Marianne, born with spina bifida, was old enough to maneuver around in her electric wheelchair. By adding an interior ramp in the mudroom, selectively removing interior walls and doors, and consolidating two small bathrooms into one larger one they created a home tailored to their needs—for a while.

There was no way to make the second-floor bedrooms accessible to Marianne. The property had no room for an elevator or stair lift, so Joanne and Peter began looking for a larger house in the same school district. Most neighboring houses had the same limitations as their own, and the idea of building from scratch gained increasing appeal. When a steeply sloped

A wraparound roof provides protection from the weather at both the car transfer area and farmer's porch. The paved courtyard formed where the two buildings come together provides accessible outdoor play space as well as parking. Garage doors are double-wide for maneuverability and extra high in anticipation of larger cars in the future.

property with a small, neglected house came on the market, they enlisted the help of LDa Architects to explore its potential for meeting their family's needs.

Joanne and Peter describe the dilemma of designing an accessible house this way: You're not planning for a child in a wheelchair, but for a future adult in a wheelchair, and a space that welcomes others to visit. You're also designing for activities that aren't performed alone, but may involve one or two more people to assist with transfers. And so hallways are just a little wider, the mudroom and bathrooms just a little larger, and furniture placement is centralized in rooms to allow ample travel space around it. After their experiences with renovation, Joanne and Peter were eager to see what could be done with a new house and a new lot.

IN THEIR OWN WORDS

"We wanted our home to be connected with the outdoors so that our entire family could move easily inside and out. Though the property is steeply pitched, the architects were able to create both a house and a yard where Marianne can navigate freely in her wheelchair. The covered deck and large screened porch give us places to be together as a family outdoors, and with paved areas beside grassy yards at both middle and lower levels, Marianne can play with her siblings and friends anywhere outside."

GET THE SITE RIGHT

Placing the house, driveway, and yard on the property was the first step. The homeowners wanted seamless integration of indoor and outdoor living areas, and so architects Doug Dick and Carter Williams located the garage near the street and put the house behind it. This layout had many benefits. Accessing the garage from the back places the driveway near the house, making a flat safe area for outdoor play. Transferring people and gear from car to house is easier when all doorways are nearby. The courtyard formed in the paved and accessible area where the garage meets the house. Tucking the house close to one side lot-line enlarges the play yard. The house footprint is long and narrow, for ample daylight and natural ventilation, and its central elevator provides easy access to both the lower backyard and second-floor bedrooms. After locating buildings on the site, the architects turned their attention to the details.

CREATING A CHILDREN-CENTERED HOUSE

Carter recalls the homeowners' request for a home with a sense of adventure and play, where Marianne could get about on her own and where all three children could explore the outdoors. A covered farmer's porch connects the garage and house, making a dry area to unload the car but also outdoor play

Floor Plan
First Floor

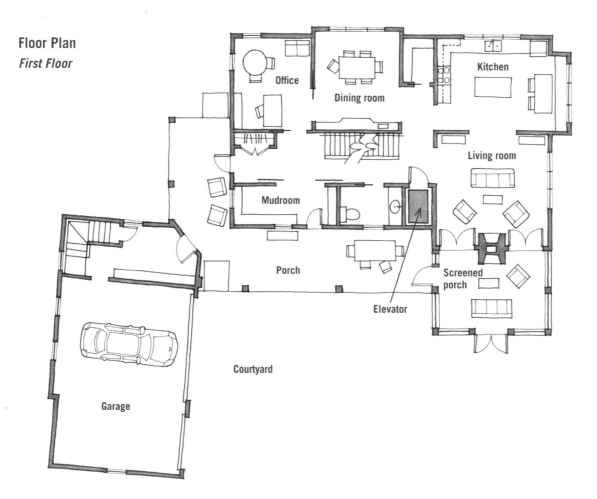

Wide aisles between perimeter counters and the central island make this a kitchen where Marianne can travel easily to prepare herself a snack.

The mudroom and casual family entrance to the left are separated from the more-formal guest entry area by custom-designed slatted doors. This way, tidiness is not mandatory for the family's backpacks, sports equipment, coats, and boots.

space in inclement weather. At the main floor a big screened porch shares a two-sided fireplace with the adjacent living room, making a continuum of gathering spaces for family games and s'mores year-round. A full basement has a family room beside the lower backyard as well as a private suite for the young woman who assists with Marianne's care.

There's a large accessible mudroom, with sliding barn doors to conceal the clutter when company arrives. In the kitchen there's a large island with plenty of maneuvering space around it, and a sturdy table near a window seat makes a good perch for keeping an eye on what's happening around the yard and house. Doors and hallways are just a little wider than required

From main house to screened porch to farmer's porch to the upper courtyard, the site offers a variety of safe ways to experience the outdoors.

A built-in window seat near the kitchen overlooks the lower backyard, and is much-used for casual eating, homework, and art projects. The trestle table has open space below for Marianne's wheelchair.

Rooms are designed to allow generous wheelchair travel routes. Wide hallways, aligned openings between rooms, and pocket doors combine to visually link rooms while giving the house a spacious quality.

by accessibility guidelines, and rooms just a little larger, allowing space for parking wheelchairs when not in use.

ROOTS AND WINGS

Safety and autonomy were guiding principles as the plans took form. Strong visual connections through the house and yard promote safety by allowing informal adult supervision. With alternative paths of travel—multiple doors to the outside, and halls and interconnected rooms inside—help arrives quickly if needed. (And a game of hide-and-seek has endless possibilities!) Site retaining walls make wide flat areas for outdoor play at upper and lower yards, and strategically placed light fixtures make it easy to get around safely after dark. A gently sloped and paved pathway around the property lets Marianne drive her chair independently between play areas. Landscape plantings line pathways to identify areas where wheelchair travel is hazardous.

Details throughout the house were chosen to promote independence and safety. Pocket and barn doors allow flexibility in room use, so that the home office can be closed for quiet homework time or opened for a sense of connection with the family. Hardware was selected so Marianne can operate

THE COST OF ACCESS

Is an accessible house more expensive to build? On a per-square-foot basis, costs are competitive with conventional construction, and the extras are offset by savings. For example, the elevator and power door operators in this house increase the overall price, but an open floor plan has fewer walls. While wider hallways and bigger bathrooms make this slightly larger than a conventional house, the architects estimate that the per-square-foot costs are comparable to building from scratch.

Double doors and a two-sided fireplace between the living room and screened porch allow these two rooms to be used in tandem for much of the year.

doors comfortably, without full use of both hands. The mudroom entry door is remotely operated from her wheelchair, and timed to close slowly so she can enter safely. Push-button controls are at each elevator landing. Bedroom lights are controlled both at the doorway and from the bed, and closet lights turn on automatically when the doors are opened. Additional electrical outlets are provided around the house for charging wheelchair batteries.

The house was designed to adapt to Marianne's evolving needs as she grows. In-wall blocking was installed behind drywall in the bathrooms for future installation of grab-bars and mobility lifts. Electrical service was provided for future installation of mobility equipment in Marianne's bedroom and bathroom. With living areas scattered throughout the house, in varying degrees of privacy and openness, this child-centered home will easily adapt to becoming a teen-centered home. ✦

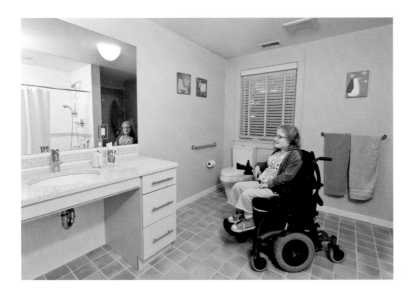

The second-floor bathroom shared by the three children has an accessible vanity and transfer space around the toilet, as well as a roll-in shower (seen in mirror).

Doorless openings flanking the fireplace create maneuvering space just where it is needed—beside the furniture groupings—and link the living room with other areas of the house.

suburban renovation

J amie was six when her parents decided it was time to stop carrying her up and down stairs. Diagnosed at birth with cerebral palsy, she used a power wheelchair for mobility and a talking board for communication. At school Jamie rolled freely through the building, but home was a different matter, where she needed help with everything from getting dressed to preparing snacks. In first grade Jamie was ready for independence and the pride of self-management. With sore backs and weak knees, Katharine and John wanted the same independence for their daughter at home.

THE HOME ENVIRONMENT

The 1930s Colonial house sits at the end of a short street abutting conservation land. This neighborhood is an eclectic mix of modest houses, set close together on small lots. With a lively commercial center and public transportation two blocks away, and a strong special-education program in the public schools, the family was committed to staying put. Katharine contacted me to see what could be done to make their home fully accessible without seeming institutional.

Their three-bedroom house was cramped for a family of five (plus Chester, a big collie). Shoes, boots, backpacks, and jackets spilled onto the hallway from a small foyer, where steps and narrow doors made passage nearly impossible. The dining room doubled as a music room and homework center. Comfy overstuffed living room furniture left little room for maneuvering in a wheelchair. On the second floor Jamie shared a bedroom with her sister, and their brother shared a room with his giant iguana, but the master bedroom was where the family gathered for bedtime stories. None of the bathrooms was large enough for a wheelchair or assisted hygiene. Jamie's three mobility devices— motorized wheelchair, manual waterproof chair, and motorized standing scooter—were signs of her mobility as well as her limitations. Two were always parked somewhere in the house.

I spent time with Jamie at home to better understand how she interacted with her environment. The parents suggested I also spend time at school, to observe Jamie in an accessible setting (see the sidebar at right). Jamie was an enthusiastic and engaging child, eager to keep up with other children but unable to coordinate her fingers for reaching and grasping objects. Her physical therapy team was optimistic that muscle control would improve with a practice that combined exercise with relaxation.

LEARNING FROM THE SCHOOL ENVIRONMENT

Jamie's school gave her a degree of independence that home couldn't provide. Special-education students are mainstreamed in classrooms with their peers, and assigned to a team of specialists. Jamie's team included an occupational therapist, a physical therapist, a speech and language specialist, a classroom aide, and a head teacher. All shared their expectations of how Jamie's muscle development and language skills were likely to progress and contributed ideas to the developing plans.

The school is designed for accessibility, with a ramped entrance, an elevator, wide hallways, child-sized accessible bathrooms, a gymnasium, and an outdoor playground. Jamie's first-grade classroom had large cushions on the floor where all children could relax for story-time, and child-sized desks and storage cubbies. The gym and school playground have plenty of space and a physical education curriculum that emphasized muscle coordination, self-control, and social skills—ideas that made their way into the design of Jamie's home.

The addition begins to the right of the chimney. An accessible family mudroom mirrors the form of the original garage to the left. Porches on two levels bring the living outdoors.

Floor Plan
Second Floor

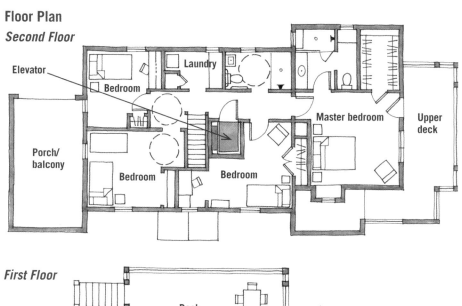

Elevator

Bedroom

Laundry

Master bedroom

Upper deck

Porch/ balcony

Bedroom

Bedroom

First Floor

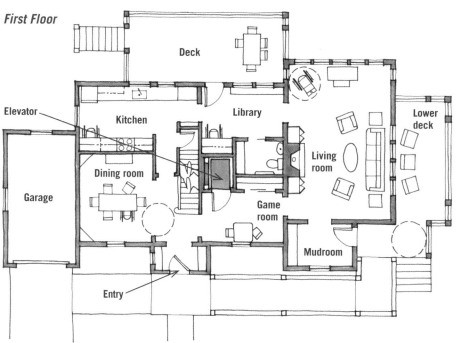

Deck

Elevator

Kitchen

Library

Lower deck

Garage

Dining room

Living room

Game room

Mudroom

Entry

First-floor hallways are wide enough to accommodate other uses. Here a home office and library link the kitchen, living room, and half-bathroom.

Wall hooks and a bench make an accessible mudroom that's easy for everyone to use. High cabinets and beadboard paneling make the four walls sturdy and functional for storage.

Locating the elevator beside the main stair and at the center of the house allows doors to be on opposite sides of the cab at alternating floors, for drive-through usage.

DESIGN CHALLENGES

The parents' initial wish list was short. They needed an elevator, accessible bathrooms, and a bedroom for each child. To save on costs for a full kitchen renovation, Katharine suggested minor modifications, such as a larger sink with a spray faucet to wash Jamie's eating tray. More space was important, but with zoning regulations that limited expansion, the house would also need extensive renovations.

The wish list grew as barriers to mobility were identified. The house had interior steps plus a three-story stairway. Doors and halls were narrow, and hardware and plumbing fixture knobs required manual dexterity. With a convoluted and narrow path of travel, Jamie would often have to back out of rooms rather than make turns. Electrical and plumbing fixtures would need updating so that light switches were within Jamie's reach and everything would be easy to operate.

DESIGN OPPORTUNITIES

A renovation project presents an opportunity to fix parts of the house that don't work, with economies of scale unavailable for smaller construction tasks. More project goals came into focus as family members' lifestyles and personalities became clear. Opportunities for improving home life were added to the wish list:

- A generous and functional mudroom for a shoe-free household
- Additional kitchen storage and countertop workspace
- A family gym stocked with physical therapy equipment
- Porches on each level for outdoor access
- Appealing places for study, both home offices and homework centers
- A serene retreat from daily chores and a place for busy parents to de-stress
- A guest suite for a live-in nanny or physical therapist

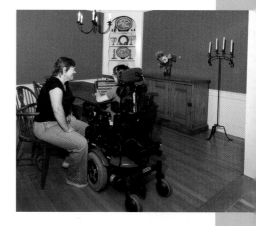

By removing the wall between the dining room and hallway, the room gains space just where it's needed.

"With the help of Deb and her team we had the opportunity to design and build our house in a way that includes our daughter Jamie as fully as possible. The modifications have made it possible for Jamie to be more independent and engaged in daily life. She is now able to explore her environment independently and play the tricks on her family of which she has always dreamed. Jamie helps with chores, including delivering clean clothes to the appropriate bedrooms. She especially loves her lowered counter in the kitchen where she can help cook. We live every day in a house that fills us with joy. We are so grateful to the entire design team for making our dream come true."

A glass-enclosed shower and wide sink vanity increase the sense of space in the master bathroom.

A wall-hung sink and low shelves increase wheelchair maneuvering space in the children's bathroom. A tilted mirror and custom cabinets create a well-appointed sink.

Jamie's lowered counter allows her to participate in kitchen activities.

Two decisions set the framework for a design where all the homeowners' goals could be met. One was to construct a three-story addition to gain a little breathing-room for the house. The other was to place the elevator at the center of the house, beside the main stairway. With these decisions, the design fell into place.

DESIGN SOLUTIONS

At 3,800 sq. ft, the house size was increased by one-third. The addition holds a large living room, with a master bedroom suite above and a gym below. Forms and materials echo the house's character, blending new areas with the old. A generous new side entrance, served by a large porch and ramp, is where backpacks and sports gear are stowed and extra wheelchairs are parked. Guests use the original foyer, where a new bench and cabinet keep things tidy.

Rooms radiate off the central elevator so that nothing seems far away. Hallways and door openings were widened and relocated to gain crucial space where it was most needed, especially in the kitchen and dining room. All doors used by Jamie are pocket or automated.

This home is family friendly, fostering independence while allowing interdependence. Children's bedrooms are equitably sized and each has unique amenities—a play loft for Tom, a sunny reading nook for Anne, and a large padded table for Jamie's bedtime stretching routine. Wide hallways at the first floor are enlivened with activity areas—a games table near the elevator and a study between living and kitchen areas. Reconfiguring the kitchen gained space for a pantry and a counter where Jamie can join in cooking.

Helping children gain independence lightens the work of parenting. Laundry appliances moved from the basement to the second floor allow all children to care for their own clothing. Bathrooms on all floors are sized for assisted personal care plus a wheelchair. In the basement there's a guest suite and family gym, with a large bathroom and whirlpool tub for relaxing tired muscles post-workout. A new master bedroom suite with a porch among the treetops makes a soothing adult retreat. In an environment where the chores of daily living can be shared by all, parents have more time to enjoy their children. ✦

a house that grows with its owner

Jenny moved across the country from Massachusetts to live in Berkeley, California, after an accident left her a paraplegic. Here she knew she would find accessible public spaces and a supportive community. The two-family house she purchased seemed a good place to live while attending law school; she would live on the first floor with rental income from housemates and tenants on the second floor. On a mostly flat street just below the famed Berkeley hills, the house had a lot of character but it was old and needed work, starting with a ramped entrance and accessible bathroom. The first step was to hire architect Catherine Roha for some basic access upgrades.

PHASE 1: A HOUSE FOR A YOUNG SINGLE WOMAN

Seven steps led from the sidewalk to the front porch, which was 20 ft. in from the sidewalk. The first challenge was finding space for a 48-ft.-long ramp.

A long ramp fits in a small yard using switchbacks. Starting to the left of the front stair, a sloped concrete ramp winds behind the hedges to arrive at a landing halfway up to the front porch, and then continues in wood to the right, arriving at the front porch.

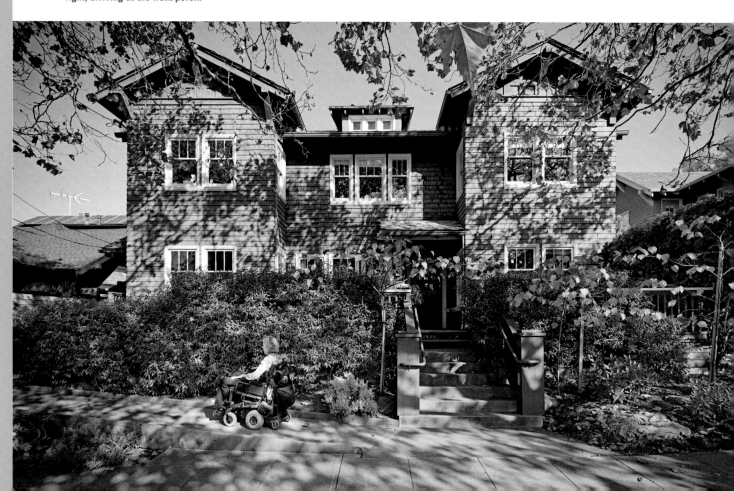

BERKELEY, A LEADER IN ACCESSIBILITY

Berkeley, California, has long been at the forefront of the accessibility movement. The first curb cuts in the nation were installed at the corner of Bancroft and Telegraph Avenue in 1970 under urging from the community. The Ed Roberts Campus in Berkeley, and its signature red ramp, offers proof that accessibility rights are civil rights and have reached mainstream status. A dynamic tenant roster includes the world's first organization run by and for people with disabilities, as well as groups involved with research, training, and advocacy to eliminate barriers to full social integration in sports and recreation, technology, employment, and family care.

Viewed from the side, the concrete entrance ramp in the front yard fits neatly between sidewalk and house. The front porch was widened as part of the project, with a low wall and hedge providing a sense of privacy from the street nearby.

Cathy's design for a gentle ramp with switchbacks fills the full front yard. A landscaped hedge between segments of ramp provides privacy from the street. Cathy widened the front porch to make a comfortably sized deck for outdoor sitting, allowing Jenny to finally enter her own house.

The original bathroom had an old claw-foot tub, completely inaccessible, and a medicine cabinet that was beyond Jenny's reach. Cathy replaced the old tub with a model that Jenny had enjoyed in rehab, with a lift-up side panel and multiple controls (see the photo on p. 67). Beside the sink she designed a storage cabinet, substituting cut-out tops for traditional drawer pulls, and adding a new tiled countertop for personal care items.

Wide openings between rooms and short hallways gave Jenny room to maneuver in her wheelchair, but the house needed some tweaking to make it more accessible. Like many old houses Jenny's had a string of little closets, pantries, and utility rooms. Cathy's plans from this period show walls removed and closets reworked with lower poles and shelves. Installing doorbells and electrical outlets within reach were simple modifications with big benefits, as were doorway changes. Between Jenny's bedroom and the bathroom Cathy added a surface-mounted sliding door, economizing on space while improving access. She placed a second new doorway leading to the stairway serving the second-floor apartment, to connect Jenny with her housemates. Cathy traded standard door hardware for lever handles, and installed swing-clear hinges to widen door openings. Built-in cabinets at the dining room and butler's pantry had heavy, sticky doors. Cathy had the drawers reworked with smooth-acting glide hardware, and replaced cabinet handles with pulls that can be operated with low hand strength.

City approvals for the project included a designated curbless parking space in front of the house for Jenny's van.

Small gestures make a big difference in the remodeled kitchen. New windows were lowered for a yard view from a seated position. The back door was widened with a 42-in. opening, and the wall separating a back hallway from the kitchen was removed for more kitchen and more maneuverability. Low cabinets and countertops are new.

BUILDING IN PHASES

Many people choose phased construction to control costs, as the work can be stretched out over several years. Others prefer to do the work in stages so that they can live at home during construction; for example, keeping one bathroom in use while the other is being upgraded. Breaking a large project into smaller parts makes the work seem less overwhelming in the aftermath of an injury, or when the course of a degenerative illness cannot be predicted. Some projects are phased to accommodate the seasons, or the availability of a particular contractor. Whatever the reasons, doing the work in smaller increments is an option that can make the difference between staying at home or having to move.

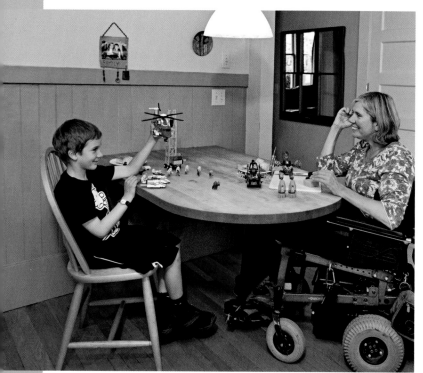

A cantilevered kitchen table, supported on a heavy-duty bracket at the wall, is fully open below, allowing Jenny to sit anywhere around it. Curved edges minimize bumping hazards and allow generous maneuvering space in the kitchen.

PHASE 2: A HOUSE FOR A YOUNG MARRIED COUPLE

Three years later, when Jenny became engaged to a man who also used a wheelchair, the house suddenly seemed small. The second floor beckoned, promising more space, daylight, and a feeling of being in the sycamore treetops to the south. Phase 2 included an elevator and various related modifications. A rear stair leading to the second floor was knocked out to make space for the elevator, just off the kitchen. Jenny contracted with cabinetmakers for a study desk in a sunny bay window on the second floor.

Site Plan

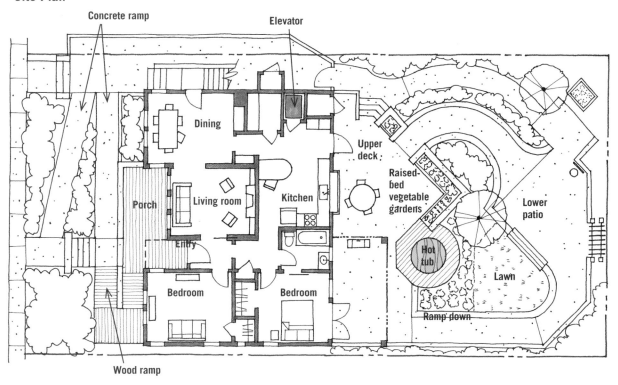

Concrete ramp
Elevator
Dining
Upper deck
Raised-bed vegetable gardens
Lower patio
Porch
Living room
Kitchen
Entry
Hot tub
Lawn
Bedroom
Bedroom
Ramp down
Wood ramp

PHASE 3: A HOUSE FOR A SINGLE MOM AND HER CHILD

A decade later and pregnant, Jenny turned once again to Cathy to make the home child-friendly. The second-floor kitchen was remodeled into a baby-changing station, equipped with a shallow sink for bathing and a long low counter for diapering. The first-floor kitchen was fully renovated for accessibility. There is a continuous countertop connecting the sink and cooktop, with generous kneespace below. Windows were replaced with larger units, mounted closer to the countertop, to give Jenny a view to the yard. Jenny chose easy-to use appliances—a refrigerator with pull-out shelves, a single-drawer dishwasher, a side-hinging wall oven. Alward Construction collaborated on the design of a table that cantilevers off the wall using a sturdy bracket, letting Jenny park her chair anywhere around the table.

The raised lawn stays soft and green, in part because it is off the main path of travel. Jenny and her son play horseshoes on the lawn, and the transfer bench allows Jenny to get out of her chair and lie on the grass.

Twenty-two years and several construction projects after moving to Berkeley, Jenny has a house that is fully accessible, but she would not have done the work any differently, even if it had all been affordable at the start. Phasing the work has allowed Jenny to fine-tune her environment to her requirements at the time. Living in the house before making changes has allowed her to see what needs work and where she likes to be at various times of day and for various activities. The house has grown with Jenny, at each stage in the life cycle. Incremental renovations are more than a way to make the work affordable; they are a way to get it right. ✦

AN ACCESSIBLE BACKYARD

Jenny could enter and live in her house, but the backyard was off-limits, except for a small deck off her bedroom. When she had saved enough money for another project, she hired landscape architect Steve Alward, who lived with a wheelchair rider and a garden lover, to design improvements. What he was able to accomplish in a small space is truly astounding. He designed a curved flagstone pathway that winds its way through the yard, ramped in two areas to connect a redwood deck with a paved patio just below. Planted areas include a small lawn where Jenny can lie on the grass and vicariously transport herself back to childhood in New England.

Raised-bed planters allow Jenny to grow vegetables near the kitchen door. There is a hot tub surrounded by a redwood transfer bench that doubles as extra seating for parties. Fragrant gardenia shrubs and climbing purple clematis cover the neighbor's garage wall, and a bamboo fence provides privacy at one side of the yard. The net result of all these features is a yard that feels rich and full of wonderful ways to be outside.

Viewed from above, this compact yard has all the features (raised-bed planters, picnic deck, hot tub, play areas, lush lawn) that one could want, linked by a gently ramped pathway.

modernist chicken coop

Marilyn and Fred knew their daughter would need assistance long after they were able to provide it themselves. Diagnosed with cerebral palsy as an infant, Megan has both mobility and cognitive limitations and requires full-time care. The family needed a house that would not only give Megan a pleasant and comfortable place to live but also make it as easy as possible for others to care for her.

Recognizing that they might also sometime need assistance themselves, Marilyn and Fred had begun considering their own requirements. Accessible features topped the list, with wide paths of level travel and a bathroom sized for at least two. The home should feel nurturing, both sunny and connected with the natural world. A low-maintenance house would give them more quality time with Megan. Constructing a new house offered a way to achieve all their goals, and when a farm subdivision came on the market with an old barn, they were attracted to the site's blend of tradition and possibility.

The farm sits on a rolling hillside with a 19th-century barn and several small outbuildings. Architect Lindsay Suter recalls walking the property and discussing what could be done there. The barn would need extensive work to make it habitable, and an elevator would restrict the already-modest living area. A derelict chicken coop on a flat portion of the site seemed the perfect place to put a new house.

Long and flat, the building is a natural expression of an interior path of travel that is simple and direct. Fortunately, the same form allows for energy-conserving design—plenty of south-facing glass with a wide roof overhang, and well-insulated north walls and roof.

A rolling barn door is a good match for the wide entrance at Megan's bedroom, and a reflection of the property's history as farmland.

COMMON SENSE AND SENSIBILITY

Lindsay sketched a single wide corridor facing south toward the barn, with rooms lined up to the north. Channeling the old chicken coop, the house is long and narrow, a fitting expression of New England simplicity but also sensible for wheelchair travel. Megan's room is at one end, a large space for sleeping and playing. Family living areas are at the other end. In between are bedrooms, bathrooms, and utility spaces.

Careful attention to details tailors the house to the family's needs. An overhead lift connects Megan's bedroom with her bathroom, where a generous shower simplifies cleanup. For hands that lack fine-motor control, plumbing controls are single-lever and closet doors have push-pull hardware. Flooring is stained and sealed concrete; embedded piping provides radiant heat in winter, while slab-on-grade construction is naturally cool to the touch in summer. Floor-level windows allow Megan to see outside while exercising on the floor.

Floor Plan

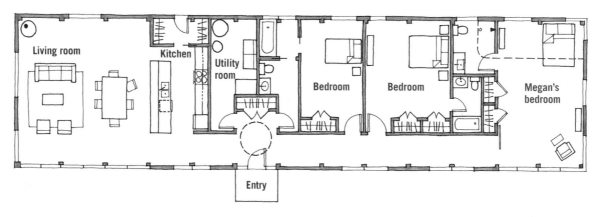

Living room

Kitchen

Utility room

Bedroom

Bedroom

Megan's bedroom

Entry

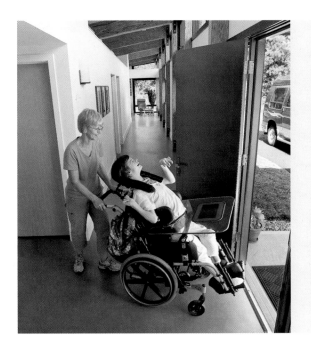

DESIGN FOR COGNITIVE LIMITATIONS

The house design has subtle cognitive aids that make it user-friendly for Megan now, and for her parents in the future. With one wall glass and the other plaster, the single main hallway makes it easier to self-orient while indoors. Room doors are brightly color-coded for way-finding: red at closets, yellow at the powder room, and blue at bedrooms. Pendant and track lighting and wall-mounted sconces avoid the need for floor lamps, whose wires can present a tripping hazard. Natural daylight streams across the main hallway and into high windows at the bedrooms to enhance indoor visibility all day.

A long straight corridor 5 ft. wide with recessed doorways gives ample space for interior travel. Brightly colored doors and big glass walls give the house a playful character but also provide "landmarks"—cues that orient people within the environment.

Marilyn and Fred wanted an energy-efficient house to keep operating costs low and make the house comfortable, so they could remain at home with Megan as long as possible. Lindsay's design includes many passive solar features, starting with a strong east–west orientation to take advantage of prevailing breezes and sun angles. The roof angle deflects strong winds and a wide overhang facing south controls sunlight, aided by customized metal louvers. Natural ventilation is enhanced by the pitch of the roof and strategic placement of operable windows. Designed for safety, ease of use, comfort, and accessibility, the house gently promotes Megan's well-being—and also that of her parents. ✦

Engineered laminated ceiling beams reduce the need for interior walls, so that living spaces can be wide open for wheelchair maneuvering.

CHAPTER 15

homes for extended families

OUR HOMES ARE NOT SIMPLY OUR OWN WHEN extended family members live nearby or visit. One designer encourages families to embrace universal design with these questions: "Do you have elderly parents? Might you someday? Do they ever visit?" Three of the families in this book have planned their homes with their extended families in mind. A Maine farmhouse is designed for the homeowner's sister (who rides a wheelchair) to visit easily and often. A house in western Massachusetts was designed to shelter two generations with shared common space. And a family in California has renovated three neighboring houses so that a man with disabilities can live with some independence while participating in the family life of his siblings and their children. Whether your family lives with you or visits, a welcoming home is one that is planned with accessibility in mind.

the visitable home

Our homes are not only where we live but also where our friends and family visit. There is a growing national trend to see our homes as places where others also have a stake in design. It is called "visitability," and consists of the idea that houses should have an accessible entrance, a powder room, and 32-in. doorways for an accessible path of interior travel (see the sidebar on p. 16). A visitable house does not require accessible bathing, cooking, or sleeping areas, but it does make it easier for people to live at home as they grow older.

Sue and Mac moved to southern Maine from Cambridge, Massachusetts, to be near extended family, including Sue's sister Alice, who had polio as a child. They wanted a home where Alice could visit freely and often. Both artists, they each wanted studio spaces where they could pursue their crafts well into retirement. The house would need to be a family-friendly place, comfortable for gatherings with their grown son and daughter and grandchildren. Sue and

A 24-ft. ramped entrance in the barn provides an accessible way to travel the vertical distance of three steps from garage to house. A low curb protects riders from rolling off the edge, and a flat landing at mid-point allows level turning space. The ramp is as much used for rolling storage carts, baby carriages, and grandchildren's wheeled toys as it is for Alice's wheelchair.

Mac chose Rob Whitten of Whitten Architects to help them design what came to be called their "Lifetime Home" in the rolling farmlands outside Portland.

Rob designed the new house around a 19th-century barn that was rescued from an adjacent site where it was slated for demolition. Rob designed the first floor of the barn as a three-car garage, woodworking shop, and ramped entrance to the house. The second-floor hayloft is used as a family room but was plumbed for future conversion to a caretaker apartment. The new house is 1½ stories tall, with guest rooms for agile family members on the second floor. Homeowner living spaces are all on the first floor. Utility areas between the barn and house (a laundry room, half-bathroom, and small home office) are linked by an art-filled hallway.

The barn at left was relocated from a nearby site and houses an interior ramp and accessible half-bathroom, as well as a garage, laundry, and woodworking shop.

Wide doorways and level floors make the house user-friendly for visitors in mobility devices, as well as comfortable for aging in place. Each room's furniture is arranged with space for a wheelchair.

Floor Plan

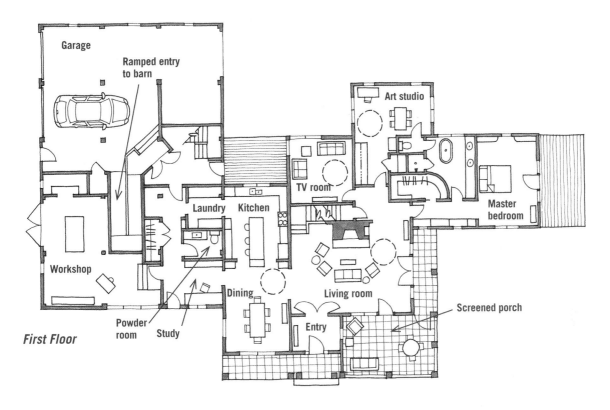

Garage

Ramped entry to barn

Art studio

TV room

Laundry   Kitchen

Workshop

Master bedroom

Dining   Living room

Screened porch

First Floor

Powder room   Study   Entry

EXPANDING THE DESIGN TEAM

Too many cooks may spoil a broth, but houses always benefit when the design process includes those who will be using the spaces frequently. Sue and Mac's house plan testifies to Alice's involvement in making sure she could visit easily. When the design phase started, Alice was using leg braces and crutches, but shortly after construction was completed she was fully reliant on a manual wheelchair. A schoolteacher by profession, Alice's gentle patience and insights helped raise the bar for design of a visitable environment. Rob acknowledges her contribution to the project's success, saying she kept the project on track with two mantras: Design for the disabled individual's specific needs, and design for their strengths.

For Alice the first thought upon entering a new environment is, "Where is there a place for me in the room?" It could be a firm chair where she can easily transfer, or a space for her mobility devices. Sue and Mac's house has the answers. Living and dining room furniture arrangements leave space for Alice's chair, both beside the fireplace and at the table. Wide pathways around furniture groupings ensure that if someone is in one line of travel, there is an alternative way for others to move about the room. Ample built-in cabinetry means that items are not left out where they might reduce travel clearances. Although the house's first floor is on one level, the sloping land just outside is from one to three steps down, and so the house has four porches where Alice can easily join the family.

The kitchen island has space below the countertop for seated meal preparation, and wide aisles around it to accommodate mobility devices. Alice uses a laptop cutting board when she helps in the kitchen.

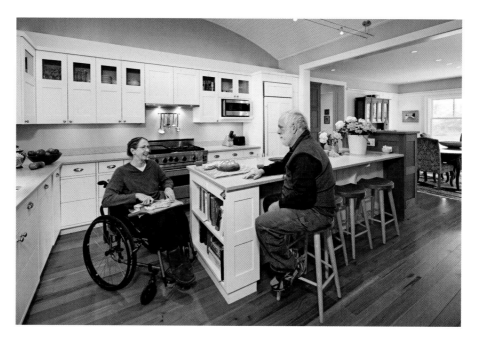

For the visitable bathroom, just inside the old barn from the house, Alice requested sturdy wall shelving rather than grab-bars, because of diminished hand strength. Counter and shelf heights were sized to meet Alice's reach range.

THE VISITABLE HOME

Specific accessible features designed for Alice's comfort also make the house easier for everyone to use:

- A wide roof over the barn entrance provides weather protection for wheelchair transfer.

- An interior ramp between the garage and house leads to an accessible entrance.

- Interior openings between rooms are wide enough for two people to pass by.

- Wide doorways connect the house with its porches.

- Cottage-style double-hung windows have a high meeting rail (above eye level).

- Lever handles or egg-shaped knobs make door hardware easy to grasp.

- The fireplace's stone hearth is set level with adjacent wood flooring.

- Sturdy side supports rather than grab-bars at the guest toilet make it easy for Alice to transfer by leaning on her lower arms.

- A fully-adjustable sink faucet at the half-bathroom puts water within easy reach.

- Open shelving for art supplies in Sue's studio makes it easy for the sisters to join in artwork projects.

- Electrical switch heights are within Alice's reach range.

- Ample built-in storage keeps floors in each activity area clear of clutter.

ORIENTING THE HOUSE IN RESPONSE TO THE SITE

Following patterns of seasonal runoff in the area, Rob located the house and barn on a slight crest and with positive natural drainage, and to the northeast end of the site so that yards, gardens, and the orchard would have full southwest sun. He set a massive fieldstone fireplace at the center of the new house, and designed living spaces that radiate off in four directions, coordinated with the sun's orientation. Indoor and outdoor living rooms lie to the sunny south, and the master bedroom catches morning sun to the east. Sue's art studio takes advantage of steady northern light. Beside the studio is a cozy library where book-filled shelves provide extra insulation in a room that only receives indirect sunlight. The dining room and a wide covered "farmer's porch" are where people naturally gather at day's end to watch the setting sun.

In the first-floor art studio, Sue and Alice enjoy making collage and painting, activities they have shared since childhood.

The barn holds Mac's woodworking studio, where floor-to-ceiling windows and nearby barn doors bring in plenty of daylight for safe use of power tools.

A wide overhanging roof above the garage doors gives all-weather protection for entering and leaving vehicles.

DESIGNING THE DETAILS

After confirming basic layout and visitability requirements, Rob turned his attention to interior details in tune with the family's preferences. Rooms have traditional styling, with wood paneling that adds durability and craftsmanship to walls. Natural wood floors add warmth at living areas, and stone tile reduces maintenance where exposed to the weather, such as the screened porch and mudroom. At the same time, an open plan with large windows gives the house a contemporary character. Single-story spaces—the kitchen and Sue's studio— have vaulted ceilings.

Being in the house conveys a feeling of spaciousness but also of comfort. With clear sightlines through the house, nothing seems far away. Rooms are defined by special features rather than walls: the fieldstone fireplace at the living room, and orchard views from the dining room. Glass doors and windows connect the screened porch with adjacent rooms.

Natural lighting and good indoor visibility will be increasingly valuable features as the homeowners grow older, as will be the ability to more easily hear people in adjacent rooms. As Rob says, "If all the doors are open, you still have a sense of privacy." This is surely a house where several generations can live together in harmony, and where the homeowners can retain their lifestyle as they age. At the same time, it's a house where Alice knows she is a welcome, fully participating member of the family. ✦

home for two families

Chris and Judy belonged to the "sandwich generation." Their three children were using the small family home in central Massachusetts as a base while taking tentative steps toward independence. Chris's widowed mother, Virginia, was still living in her multistory Brookline house and had taken a few bad falls. Hoping to help his mother settle into a new home before an emergency forced her to make poor choices, Chris and Judy made an important decision. They would find a place large enough for the whole extended family and with qualities that they all wanted: a natural setting, near a town, and accessible.

Chris was confined to a wheelchair since a wrestling injury in high school left him a paraplegic. Active in the disability community, he had friends and colleagues who used mobility equipment of all sorts. Judy was a respiratory therapist whose work transporting patients had caused various injuries to her back, knees, and feet, requiring multiple surgeries with occasional assistive devices. Virginia had severe stenosis and had gone from walking short distances to using a walker and then a wheelchair in seemingly no time.

Design of the sunroom evokes the American southwest, where Chris and Judy have fond memories, but it is also a natural expression of the house's environmental controls. South-facing windows collect sunlight while a tiled floor holds its heat.

195

Floor Plan
First Floor

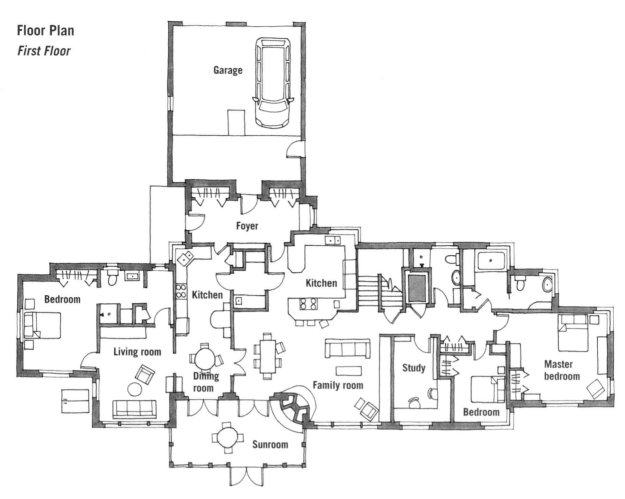

Garage

Foyer

Kitchen

Kitchen

Bedroom

Living room

Dining room

Family room

Study

Master bedroom

Bedroom

Sunroom

A shared sunroom and patio connects both apartments with each other and with the yard outside. Virginia's apartment is to the left, while Chris and Judy's is to the right.

Architect Bill Austin was a colleague whose projects were often reviewed by local access boards on which Chris served. The two men had developed a shared vision of the ways that accessible design can improve lives, as well as a deep friendship. Meeting with Bill was the starting point for Chris and Judy in exploring ways to create a home they could share with Virginia, and where their grown children, and eventually grandchildren, would always be welcome. They were searching for houses in the Berkshires that might meet all their requirements when a property came on the market that offered a clean slate.

ONE HOUSE, TWO HOMES

The idea of a "modified duplex"—one large home with a separate in-law apartment—took hold early in the design process. Chris sketched a diagram of a common service area with private living space on either side. By bringing an underlying structural order to the diagram, Bill was able to translate it smoothly into detailed floor plans. The layout has a common entry with a mudroom and garage (see the floor plan above). Living rooms share a common wall, with frosted-glass doors that function as a modern-day smoke-signal: Open doors mean visitors are welcome. A shared sunroom connects the apartments with each other and with gardens to the south.

The large kitchen is a direct expression of Chris and Judy's love of entertaining. A peninsula centered in the living space is a showpiece for Chris's culinary talents and a place where he can prepare meals and stay involved with activities all around him. His work surface is 30-in. high and holds the cooktop; Judy's workspace is 36 in. and houses the dishwasher. Appliances used by both are in the center.

Virginia's apartment reflects her needs as a woman newly transported from city life. The layout is compact, keeping travel distances short between rooms. With a few beloved pieces of furniture, she wanted a home with traditional styling. The kitchen is compact and the bathroom has a raised toilet seat and a grab-bar built into adjacent shelves. Good lighting was a high priority for this avid reader, as was locating the bedroom where she could see neighbors.

Chris and Judy have the kind of household that attracts people to stay, both overnight and long term, and their apartment reflects their expansive and exuberant personalities. The heart of the house is a spacious family room consisting of living, dining, and kitchen areas. A bedroom wing on the quiet end of the house has two east-facing bedrooms, including a master suite with a large soaking tub plus Chris's office. An open second-floor loft holds another guest room and open play space. From the start Chris and Judy wanted a house large enough for all three children to be there together, eventually with their own children, partners, and friends. While the house has ample space

Concrete fireplaces in the house have hearths set to Chris's wheelchair height for easy reaching. Bookshelves and a television set are placed where they can be seen throughout the house's main living areas.

CREATING A STABLE INDOOR ENVIRONMENT

The first step is to create an environment that can maintain a comfortable temperature for long periods of time, regardless of the weather or the sun's angles (no small task in New England). Walls built with insulating concrete forms hold heat in winter, deflect it in summer, and eliminate drafts, and roof areas likewise have extra insulation. Windows and roof overhangs are sized and located to introduce warming winter sunlight and cool summer shade. Dark floor tiles and thick concrete fireplaces hold solar warmth and the fire's heat, radiating it back into living spaces. In the end it's the sunroom, acting as a solar collector, that allows Chris and Judy to fine-tune indoor air temperatures, simply by opening or closing the doors.

Beveled edges seem to enlarge wall openings to make the most of modest window sizes in a superinsulated wall.

for overnight guests, most of the living takes place in the family room, which seems comfortably sized for either a couple or a crowd.

INDOOR CLIMATE CONTROL

People with spinal cord injury have extreme sensitivity to temperature changes, and a primary challenge in design is to create a consistent indoor environment where cycles of heating and cooling are nearly imperceptible. Spinal injuries prevent the body from thermally self-balancing, a problem that Chris experiences, and one which increased as Virginia's age advanced. Poor circulation in cold weather prevents the feet and hands from staying warm through movement. A person might need to pull a sweater off and on several times to find a personal comfort zone. In hot weather, self-regulation becomes an even larger problem, and Chris speaks of nearly fainting when his body temperature once rose to 102°F. Humidity adds to the challenges of staying comfortable in summer, as the injured body does not perspire to cool itself.

Indoor temperature and humidity control is based on the principles of passive solar design (see the sidebar above), along with mechanical equipment that fine-tunes indoor air temperatures. In-floor radiant-heating tubes keep room temperatures uniform—unlike traditional systems with radiators that overheat certain areas while leaving others cold. High-velocity air conditioning with mini-ducts spaced regularly around the house provides dehumidification and uniform cooling, without the blasts of cold air from a typical system with fewer but larger ducts. Ceiling fans gently send warm air back to living areas. Rooms are neither overly dry nor humid, a factor that causes visitors to say they've never felt so healthy. Although the special environmental features increased construction costs slightly, it was a small price to pay for Chris's year-round comfort. ✦

family compound

T here is a big difference between making a home ADA-accessible and making a home personally accessible: One is about getting into the house and the other is about quality of life, and a home with full accessibility has both. This is the approach taken by a cohesive California family whose members have created an extraordinary setting in the wake of tragedy. Their story starts with an automobile accident that left 16-year-old Douglas a paraplegic with TBI (Traumatic Brain Injury), unable to perform many of the activities of daily living without assistance. It continues, 30 years later, in a small town south of San Francisco, where three out of four grown siblings live as neighbors. The common thread to the story is a family's love and commitment to each other.

Adjacent properties are connected by paved pathways and ramps, for a fully accessible set of yard areas. A shared fence helps make this arrangement work, as the sliding door can be closed when one party wants a little privacy.

Douglas lives in the fully accessible off-white house with his aunt in an apartment upstairs. His brother Mark lives in the visitable red house next door. This extended family has created an environment where Douglas has all that he needs within reach.

THREE HOUSES, ONE FAMILY

On a quiet street there are three houses with accessible entrances. One house (the off-white one) has a three-way ramp leading to the front porch 18 in. above the sidewalk level. Another (the red one) has a platform-type wheelchair lift attached to the front porch 30 in. above the sidewalk. At the third, across the street, the front yard has been re-graded to bring the land level with the front porch 7 in. above the sidewalk.

Douglas lives in the first house, with his Aunt Susie as caregiver in an apartment upstairs. Brother Mark lives next door with his wife and young daughter, and sister Evi-Lynn shares the third, across the street, with her husband. Parents Bob and Evi live a short drive away, as do Douglas's brother Scott and his family. Here is how this unique arrangement came to be.

It took ten years for Bob and Evi to realize their dream of helping their grown son live independently. They searched for a neighborhood where the land is flat, within a few blocks of downtown and served by accessible public transportation. The property they found comprised five separate parcels of land with four small houses, just the size for their extended family and with room to grow. Most promising for their purposes was a dilapidated house, a teardown really, straddling the boundary between two adjacent lots. The family hired architect Bill Bocook to see what could be done. Navigating town approvals, from historic to zoning reviews, added time to the planning process, but it was worth the wait. The web of homes and yards they have created is more than a good home for Douglas—it is a way for this close-knit family to stay connected through all phases of the life cycle.

ADDING VALUE BY MOVING A HOUSE

The footprint of Douglas's house was large enough for a good-size living area, and the lot was large enough for a new bedroom wing. Because the building sat on the line between two lots, it would need to be moved to a new foundation to keep property development options open. This was an opportunity both

Site Plan
First Floor

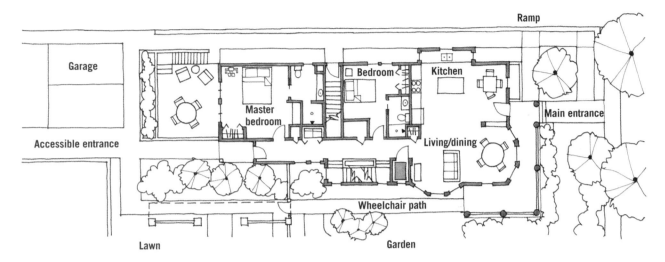

Garage

Accessible entrance

Ramp

Master bedroom

Bedroom

Kitchen

Main entrance

Living/dining

Wheelchair path

Lawn

Garden

A paved pathway swirls through the side yard so that Douglas can easily visit his favorite flowers. This yard is a separate buildable lot, where Douglas's parents plan to construct their own accessible home for aging in place.

to create a habitable basement, with large windows beside outdoor lightwells and with high ceilings, as well as to place the first floor closer to ground level and shorten ramps on the exterior. It was also an opportunity to create an accessible yard, big enough to give Douglas various ways of being outdoors.

Interior walls and corridors in the old house were removed to enlarge each room (see the floor plan above). A living/dining area shares the first floor with a kitchen/breakfast area. The house had two stairways, one up, one down. These were eliminated in favor of a new U-shaped stair connecting all three floors, beside a new elevator. The first-floor plan gives Douglas a new bedroom with adjacent bathroom, along with a half-bath for guests and a laundry room. Each entrance has a ramp leading to ground level, and some entrances also have steps.

Interior walls in the house were removed to make a more open plan, with dining and living areas linked to the kitchen and breakfast corner. Low walls and cabinets provide places to store and display items, so that floors are kept clear for safe travel in mobility devices.

A window in the cab, and corresponding windows on the house's wall, make an elevator where Douglas does not feel claustrophobic.

The house's showplace is the basement, where a large-screen television and comfortable lounge chairs draw the family together to watch their favorite hockey team, the Sharks. An adjacent kitchenette and dining table make this a well-used part of the family compound. The basement also has a smaller multipurpose space designed as a workout room and a fully accessible bathroom. Outdoor lightwells bring daylight through windows in foundation walls, and sliding doors to a sunken patio help keep this level from feeling cavelike.

Behind the house is a freestanding garage with battery-charging stations for Douglas's motorized mobility equipment (see the sidebar on p. 202). Beside the house is the family park, where paved pathways weave through gardens of roses, Douglas's favorite flowers. A wooden fence separates the gardens from a lush lawn, partially paved using a perforated concrete block that allows grass to grow through it. The block creates a wheelchair-friendly yard that never gets muddy.

VISITABLE NEIGHBORS AND SIBLINGS

Mark is the family cook, and his house next door is designed for accessible visiting. In the kitchen is a large island with an attached composite-stone countertop, open below for wheelchair seating on three sides. The first-floor bathroom has an accessible toilet and sink, and interior doorways have all been widened for extra maneuverability, whether in a crowd or a mobility device. Douglas enters the house using a vertical platform lift at the front porch, level with the first floor. In the backyard a ramp connects a large open deck to the ground level. Compacted stone dust makes a firm and level surface for pathways that meander through the yard, giving Douglas full access to

The basement theater is where the family gathers for sports and games. Stepped risers and reclining lounge chairs make every seat the "best seat in the house." Douglas prefers a firm stationary chair with adjustable footrest, where he can transfer easily from his wheelchair.

The lawn beside Douglas's house has a portion designed for easy rolling. Permeable concrete pavers have holes where grass is planted, keeping the lawn level and seldom muddy. Thin paved pathways through the lawn designate where the pavers are located, for safety in any future excavation and for service to the yard's irrigation system.

TECHNOLOGY FOR ACCESS

Grab-bars in the garage corner where his wheelchair batteries are charged allow Douglas to transfer more easily from one power-chair to another. The transfer station requires space for two wheelchairs, plus travel and maneuvering space. A generator installed outside the main house provides backup electricity for the elevator, sump-pump, and basement, so the whole family can gather there in the event of a power outage.

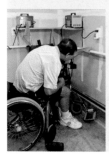    

outdoor activities. Raised-bed planters in the backyard enable him to help with gardening, or play hide-and-seek with his niece.

Evi-Lynn lives with her husband Colin across the street in a bungalow set close to the ground. Visitability upgrades at her house are a work in progress; the first project was to raise the ground level at the front yard to make a gentle ramp leading from sidewalk to entrance. The compound's fourth building is a small two-story house beside Mark's, currently being used to generate rental income, which offsets expenses at the compound.

The landscaped yard beside Douglas's house is the site for a future new home where Bob and Evi can be closer to their growing family. Bill Bocook is starting to work with them to design an accessible home where they can live out their years, surrounded by the family that they have raised to care for each other. Visiting the family during the preparation of this book, it was clear that they have created an environment where the physical setting reflects emotional ties between people—a balance between freedom and interdependence, close without being intrusive. It seems a healthy and happy way for all to live. ✦

The front porch is wide enough to accommodate not only the lift, but also a wheelchair turning space and room to move out of the way as the lift door is opening. Porch railings were raised to the level of lift walls using metal fencing, a thoughtful detail that keeps the lift from seeming like an afterthought.

Mark's house is made accessible by means of a platform-type lift built into the front porch. A paved pathway from the sidewalk makes a smooth rolling surface to reach the lift.

Mark's kitchen has an accessible table built into the island for casual family meals and aisles around the island for wheelchair travel. A pass-through window between the kitchen and dining room beyond limits the distance traveled in serving meals, which allows Douglas to help at family dinners.

CHAPTER 16

aging in place

STAYING AT HOME THROUGH THE LATER YEARS is a dream shared by most people. An accessible environment makes it possible to live with independence throughout the life cycle. The designer's challenge is to incorporate features that improve safety and comfort throughout the home. Of course, we never know what kinds of disabilities may be in store for each person, and so a responsive design is one that makes it a little easier to see, to hear, to find our way, to open and shut doors and operate plumbing fixtures, to reach light fixtures and controls, to move about—in short, the Accessible Home.

reconstructed barn

J an was fit and healthy, but she knew it was prudent to plan for a time when she might need help, or have friends in wheelchairs. Newly widowed, she was especially attuned to the hazards of living alone, and wanted a home where she could grow old in comfort and safety, but the idea of moving away from the old house she and her husband Robby had remodeled together had little appeal. Planning a new home would have to wait until she knew where and what she wanted and had put her personal finances in order.

The family homestead was a few blocks from the town center in rural Connecticut, with an old barn that had captured Robby's imagination when the young couple first saw the property. Robby had sketched out many versions of how the barn might be converted to living space, but the town required special approvals and the outcome was uncertain. Restoring the barn would have to wait until work at the main house was completed, and Robby passed away before they could carry out their ideas.

The new barn is a replica of a Victorian structure that stood on the same site.

Tall ceilings and a lack of interior walls give the home a feeling of space, while the open plan makes it easier to communicate with someone in the next room.

Zoning requirements that the building match the original footprint precluded adding a porch extension, but a recessed porch turned out to be a fitting alternative. Furniture and plantings make a welcoming entrance as well as a comfortable outdoor room.

Floor Plan

First Floor

- Elevator
- Office/ art studio
- Garage
- Den
- Entry porch
- Workshop

Second Floor

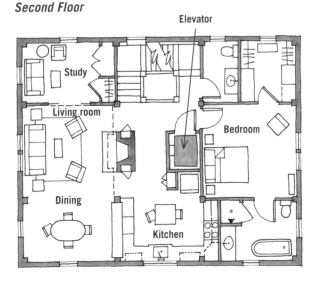

- Elevator
- Study
- Living room
- Bedroom
- Dining
- Kitchen

During this time their son Russell had completed design school and was establishing an architectural firm with his fiancé Mary Jo. At family gatherings Jan's next home was a hot topic, but neither was ready to make the necessary commitments. Russ now sees this as a time when he was developing the technical and business expertise to handle a complex project like restoring the barn, along with the confidence and interpersonal skills to work together with Jan as equals.

A FAMILY AFFAIR

The barn's renovation owes much to serendipity. The structure was starting to crumble just as mother and son were ready to enter a professional contract. Ross felt he had the judgment to advise her well, knowing that this would be a big investment, financially and emotionally. They applied to the town for approvals to convert the barn to a residence, and started the design process.

Basic layout decisions were dictated by the shape of the building and Jan's needs. The inclusion of an elevator allowed the placement of living space and a bedroom on the second floor, where being surrounded by treetops conveyed

By carrying floor and wall materials from adjacent rooms into the cab, the elevator becomes an integral part of the house. Transparent cab doors keep the elevator from feeling claustrophobic.

IN THEIR OWN WORDS

"My favorite aspect of my home is that I can spend most of my life on my second floor, looking out at and into the surrounding trees, and sharing my vantage with the birds! I love that my 'lofty living' allows my windows to be free of privacy shades or curtains and that my spaces are filled with light, so that I can absorb every bit of the daylight and sunshine."

A small den beside the entrance doubles as a cozy living room or a waiting area beside the home office.

Frosted glass doors bring light throughout the first floor. Pocket doors allow this area to function as two small rooms or one larger space.

a sense of privacy. Service areas would be on the first floor: garage, laundry room, office, and art studio. A guest suite would be tucked under the eaves on the third floor—good for visitors now, and possibly a live-in caregiver in the future. The barn's footprint was small but it was adequate, and with mostly single-story living, Jan's new home would be easy to maintain.

Renovation projects have their inevitable challenges. About this time, a restoration firm was brought in to evaluate the structure. Many of the old timbers had dry rot and insect damage, but some materials seemed salvageable. Russ went back to the town with the news. The zoning board would accept a building that replicated the old barn, with the same shape, same location, same appearance, and same heavy timber construction with wood bracing and traditional pegged fastening. The project was back on track, a new building nearly like the old.

Design details came together quickly. Safety features would include nonslip floors, good lighting, and zero-step showers to reduce the chances of falling. Jan wanted a kitchen that would be easy to use from a seated position, with lowered countertop areas, pull-out shelves, an appliance hutch, and plenty of counter space to reduce effort and keep clutter at bay. She hoped to reduce operating expenses with programmable lighting, a well-insulated house, and energy-efficient appliances.

A NEW OLD HOUSE

Jan had lived in older houses all her life, and was concerned that a new building wouldn't feel like home. No problem, said Russ. The design has traditional elements like turned wood balusters, two-part interior casings, face-framed kitchen cabinets, and a claw-foot bathtub. Jan chose colors schemes and finishes with a strong Colonial character—deep green and lighter sage paints, bubble-glass lighting pendants, and a wrought-iron chandelier over the dining room table. And the exposed post-and-beam timbers give the barn instant authenticity.

Living space in the barn would be smaller than the house where she'd raised her family. Where would she put her furniture? What could she part with? Jan was concerned the new home would feel tiny. Again, Russ had answers: multi-use rooms. On the first floor, one enters through the porch into a welcoming entrance room that doubles as a cozy living room or waiting area. Sliding glass doors lead to an adjacent study, which doubles as a sewing room and art studio. Vases are stored in the laundry: With a broad countertop and sink, it works as well for arranging cut flowers as for washing and folding clothing. On the second floor, the library doubles as a guest room.

An island table makes for multilevel work surfaces and seated meal preparation. High windows and translucent window coverings ensure even daylighting.

Countertops at varying heights (including a pull-out shelf) create options for preparing and serving meals. The low wall connecting kitchen and dining areas increases the feeling of space in both rooms, and makes a handy serving counter.

Wide doorways, strategically placed, simplify travel between bedroom and bathroom.

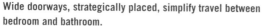

In the powder room, wood paneling installed over plywood backup will allow grab-bars to be placed where they're most needed in the future. Color contrasts improve visibility within rooms. The wood chair rail and beadboard wainscot allow those with low vision to navigate within the room using touch.

A sense of space can go a long way toward making small houses feel large, and with 1,200 sq. ft. per floor, the barn is no exception. Wide doorways and high ceilings keep rooms from feeling tight. Open plans mean more usable space, without wall thicknesses and duplicated paths of travel between rooms. Large windows blur the separation between indoors and out. Pocket doors expand smaller rooms and give intimacy to larger areas.

Efficient storage—plentiful and scattered throughout the home—adds space by reducing clutter. In the barn, utility closets tuck into alcoves beside the fireplace. A coat closet fits under the main stairway. Eave crawlspace becomes an attic, reachable from the stairway. Like the cabin of a boat, with a place for everything and everything in its place, there is always enough room.

PARTY TIME

When construction was finished, there was a big celebration. Jan estimates that 150 people came to the party. People stood in hallways and sat on the stairs. The pass-through between kitchen and dining area made light work of serving and cleanup. As she recalled the many ways her children and friends had helped with the barn conversion and construction, and Robby's original vision for developing the property, she knew she was finally home. ✦

aging in place without adding space

As Dorothy found in trying to return home to her Kansas prairie from Oz, sometimes the answers to our questions are right in our own backyard. Roberta and Ed wanted a master bedroom at the first floor of their Illinois farmhouse so that they could live comfortably at home in retirement. All rooms in the house were heavily used, the first floor as shared living spaces and the second dedicated to sleeping areas. A separate guest cottage was linked to the house with a step-down hallway. An addition seemed the only way to gain space for a new master bedroom suite, so the homeowners entered into a design contract with architect Chip Rorem of Ralph Rorem Architect Ltd.

Windrose Farm is a purebred Arabian horse farm dating from the 1850s, constructed over time as a series of interconnected structures sided in clapboards and cut stone. The enclosed hallway to the guest cottage was the newest structure, built in the 1980s. Roberta and Ed knew the addition would have to be historically appropriate and well integrated with the house's character. Chip took one look at the property and saw a graceful solution, one that avoided building an addition. He suggested reversing the first-floor guest area with the current second-floor master bedroom, and the project was launched.

The one-story link connects the main house with the new master suite (at right) while maintaining the house's mid-19th-century spirit.

The enclosed hallway from the main house to the master suite was ramped rather than stepped for an accessible path of travel, and widened to accommodate bookshelves on either side.

LUCK MEETS GOOD DESIGN

As with most good ideas, serendipities abounded on the path from design to implementation. The guest cottage was simply remodeled to a master suite. The connecting hallway was just long enough, and the floor-level changes just high enough, that the floor could be comfortably ramped, with a 1:12 slope (1 ft. vertical distance for every 12 ft. horizontal) and with intermediate flat landings spaced evenly along the hall. When Chip suggested adding meaning to the hallway by repurposing it as a library, the homeowners were delighted with extra space for storage and display. Chip reconfigured the connector with its exterior walls pushed out 12 in. on each side and lined with bookshelves, alternating with windows for daylight and views.

Floor Plan
First Floor

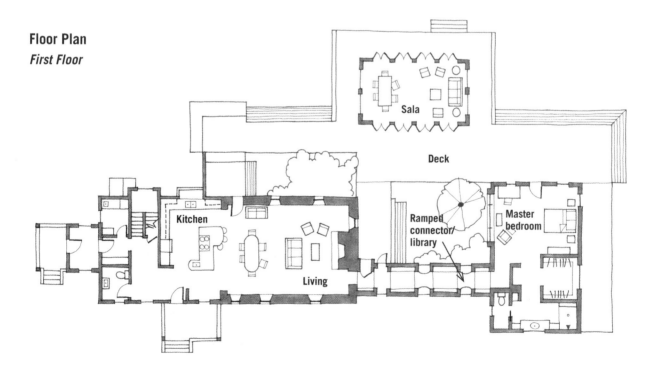

OUTDOOR ROOMS

Roberta and Ed knew they wanted to be able to step outside easily from their new bedroom, and envisioned a small deck with steps down to grade. With three doors leading to the backyard, Chip suggested linking these with a deck that wrapped around the rear of the house and with gentle steps at level changes. The new deck design accommodates ramps at each end and at the new hall door, tabled for now but ready to be added when needed.

The project also includes a freestanding screened porch. Roberta and Ed had enjoyed covered outdoor rooms called "salas" during travels in Thailand, and wanted their deck planned with a future sala in mind. They commissioned landscape architect Astrid Haryati to design an Asian-inspired structure with screens, in response to the Illinois climate and insect population, and level with the decks. A focal point for each of the four building wings and at the heart of the homestead, the sala is an accessible outdoor space where the homeowners can live out their years at home.

Windrose Farm is a success story in many ways. It's an example of how smart planning can reduce construction costs, and the evolving needs of homeowners can be met with remodeling rather than building anew. For Roberta and Ed, reconfiguring rooms they already owned left space on the property, and in their budget, to create the kind of rooms that made the house truly special—not only accessible but filled with happy memories and ready to create new ones. ✦

The new sala is linked to the house with a wraparound teak deck level with the new master bedroom (beyond at right) for an accessible path of travel around the house and site.

With its back door fronting on the community driveway, the house has a welcoming presence for visitors who arrive by foot or wheelchair. Outdoor decks at the second and third floor are reachable via elevator. The paved patio at ground level adds a third way to be outside.

accessible tower

Most people think of single-story living when they think of accessible homes, but not Rick and Marta. The couple had hoped to find a home on Seattle's Whidbey Island with an ocean view, but given the high cost of waterfront property they looked inland, and upward. Rick has MS, a disease of the central nervous system affecting the brain and spinal cord, which has caused a slow decline in his abilities over two decades. Even though he could no longer walk the craggy shores of Puget Sound, he had high hopes of living where he could enjoy the ocean. This would require a hilly setting—or an elevator, something they were unlikely to find in a neighborhood of modest older homes.

Their hunt led them to a cluster of new houses just starting construction. The developer gave them the good news—a small lot on the crest of a hill, approved for a two-story house, was available; and the bad—it was not accessible. He suggested a meeting with the architect, Ross Chapin, to see what could be done. Because construction had not yet started, Ross was able to reconfigure the interior layout and tailor the design to Rick's needs.

Floor Plan

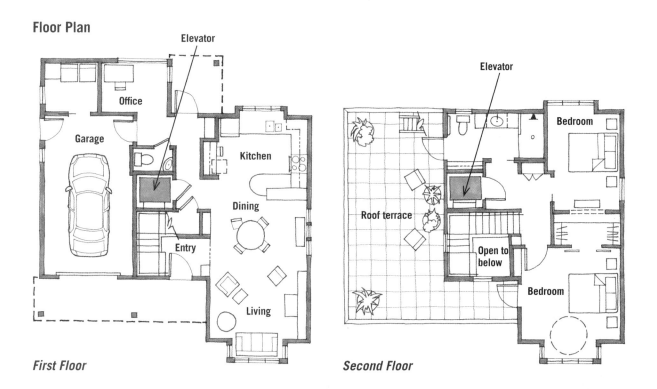

First Floor

Second Floor

By locating built-in cabinets, shelving, and fireplace against the exterior wall at left, all circulation—from aisles between furniture to stairs and elevator—is centered in the house for efficient use of space.

Without changing the house's overall square-footage or footprint, both of which had been approved by the town, Ross was able to re-design the floor plan to include an elevator. He did so by increasing the house height to three stories, and by lowering ceilings to 8 ft. to stay within zoning height allowances.

LIVING WELL, LIVING SMALL

Rick and Marta's house is evidence that an accessible home doesn't have to be extravagant or large, just thoughtfully planned. Finished living areas comprise less than 1,500 sq. ft, not counting the garage and elevator, yet Ross's design creates many places to be, both indoors and out, to give the house a complexity that is unusual in buildings this size. The house has many amenities and a spacious feeling, thanks to an abundance of natural daylight and the fact that all rooms have windows in at least two directions.

On the first floor there is an open living/dining area with a kitchen at one end and a fireplace at the other. The single-car garage is approximately 5 ft. wider than a public parking stall, but with a covered porch extending along the front entrance, wheelchair transfer to and from the car can take place outdoors if needed. Utility spaces—a powder room, mudroom, and a combined laundry/mechanical room—are compact and functional, located along a service hallway off the garage that includes a pocket door leading to a small study.

The second floor has two modest-size bedrooms and the house's only accessible bathroom, a generous space with a roll-in shower and a door leading to the deck outside. A closet between the bedrooms was designed as a piece of built-in furniture, fitting snugly between floor, ceiling, and wall surfaces. It is easily removable if the homeowners opt for a single large bedroom. Pocket

Beside the kitchen and in the mudroom, the elevator is at the center of the house and easy to reach from any room. An accordion grille makes a transparent cab door, which stacks neatly to one side in the open position.

An open deck at the second floor provides an outdoor pathway between the bathroom and master bedroom.

An elevator (interior at right) and sliding door make the third-floor deck fully accessible and complete the house's 360-degree panorama views.

Rick's third-floor pavilion allows him views of the wider neighborhood to add a sense of space to a small home. The shared driveway and lush landscaping keep neighborhood homes within friendly reach but with a degree of privacy.

doors and an open stairwell keep the second floor from feeling crowded, as does a roof terrace over the garage.

The third floor is a glorious three-season pavilion, Rick's personal retreat and reachable from inside the house solely by elevator. Only 110 sq. ft, it has 360-degree views and a small roof terrace. Able-bodies climb an outdoor alternating-tread stairway from the second floor deck to visit Rick in his aerie, unless invited to share his elevator.

The front entrance shares a driveway with the garage. Seattle's wet weather makes a covered entrance a must-have. Here, the second-floor balcony creates a dry place to transfer from car to house, with either a side-access or a rear-access vehicle.

TAILORING A HOME TO ITS OWNERS

Everything takes more time in a wheelchair, and the house abounds with responsive details. For example, arriving home on a wet Seattle day can be messy business, pausing out of the rain to shake off an umbrella, unlock the door, and remove a raincoat. With a 5-ft. covered landing at both front and back entrances, and a short hallway that connects living spaces with the garage, the house has a welcoming presence, whether arriving by car, foot, or wheels.

Considering the various tasks used in preparing meals led to inspired kitchen details. A low wall makes a serving counter between the kitchen and dining room, incorporating an appliance "garage," a cabinet where the toaster, microwave, and coffeemaker are parked. The floor of this cabinet is level with a low countertop, just the right height for Rick to use while seated in his chair and with knee clearance below. Dishes, flatware, and cereals are stored in drawers nearby, making breakfast a breeze rather than a chore.

In the bathroom, accessible features ensure low maintenance. The floor slopes imperceptibly toward a drain at the shower, creating one curb-free surface with fewer corners where dirt can collect, and making it easy for Rick to transfer to his shower chair. By locating the showerhead and controls on a short wall facing away from the bathroom, the design avoids glass and the need to squeegee it dry. Bathroom sheathing is plywood, which will make it easy to install grab-bars over wall tile in the future.

SMART CHOICES, SMART DETAILS

Marta and Rick visited showrooms to find appliances that seemed a good fit with Rick's abilities. They chose a refrigerator with an upper door that opens above Rick's knees so that he can pull right up to the fridge to take out food. The freezer occupies two lower drawers, for easy reach with minimal shuffling of objects. They chose a shallow single-drawer dishwasher that Rick can use without bending down or dropping items. In the living room, the gas fireplace can be turned on using either a timer or an override switch.

Built-in features throughout the house were designed for safety. Sealed and stained concrete with in-floor heating makes a sturdy, uniform, and level walking surface at the first floor. The second floor is a warm wood with a nonslip finish, and Rick's pavilion has nonskid concrete pavers. Stairway treads are hardwood, edged in a contrasting rosewood band, to make it easier to see level-changes. Rocking toggle switches at fingertip height make it easy to control light levels for safety and visibility. Step-lights installed at hallways and decks mark a nighttime path of travel. Built-in storage keeps floors and work surfaces clear for safe travel around the house.

Rick and Marta enjoy their garden bench beside the patio. In lieu of a fence, and with seating on either side of a shared backrest, it is a neighborly alternative to a solid wall separating yards. A stained-glass wall panel adds a sense of connection to surroundings while maintaining privacy at each side.

OUTDOOR LIVING

When mobility is limited, the senses of sight and sound heighten our experience of the world, and sitting in a pleasant setting takes on greater importance. This cottage provides many wonderful ways to enjoy the outdoors. A bay window beside the gas fireplace on the first floor makes a cozy place to watch birds in the shrubbery outside. Outdoor decks allow for stargazing and wildlife observation, with deer frequent visitors from the nearby forest. The third-floor pavilion is a great place to experience Seattle's sparkling night skyline. As the landscaping was being completed, Ross designed a garden seat with a low backrest shared by adjacent properties—an antidote to the "keep your neighbor out" fences so often used when homes are close together. From the landscaped patios to the rooftop decks, Rick and Marta have the best of land, sea, and air, and a house that brings nature's wonders to their doorstep. ✦

prefab and modular

Many architects are exploring the design potential of factory-built housing, a low-cost alternative to conventional construction. In the carefully controlled indoor environment of a factory, building parts can be assembled year-round without weather-induced delays. And when the house is delivered to the site, either as a whole or in parts, the final assembly goes quickly, saving time and money.

The idea of prefabricated construction is not new, but increasingly it appears an attractive option for situations unique to our times. As we have seen in Louisiana, affordable portable housing can help communities rebuild quickly following a natural disaster. And in Washington State, Accessory Dwelling Units are being marketed as small apartments attached to a house as an in-law flat. Neither model is designed to be accessible on a large scale—yet.

In this final chapter we will look at two very different pre-fabricated homes. Both are modular, meaning that the final assembly on site involves connecting parts ("modules") together and completing the work with a construction crew.

Beau's simple, angular lines seems right at home among the rugged hills of eastern Washington state.

FabCab started out as an exhibit for accessible modular construction at the Seattle Home Show.

Floor Plan

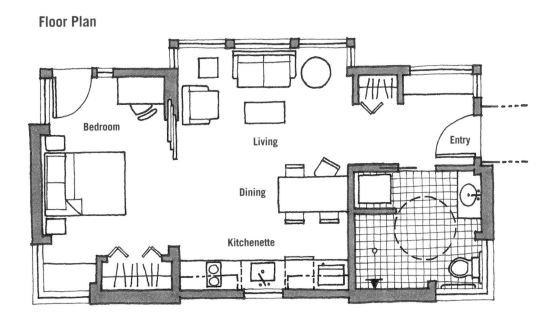

And both are designed to be accessible. Differences between the two houses reflect the regional cultures, as well as the creative energies of two design teams.

A WASHINGTON PROTOTYPE

This tiny house situated in the high desert of eastern Washington began as an exhibit at the Seattle Home Show. Architect Emory Baldwin saw a growing gulf between what people want and the choices they have. Emory set himself the challenge of designing an accessible modular house, with one catch: the accessible features had to be entirely invisible to have broad appeal – equally successful as a bachelor pad or nanny suite. He called the design FabCab.

At 550 sq. ft. the house is tiny, but with careful planning it has all the amenities of a larger home. There is a fully accessible bathroom and compact kitchen. Sliding interior doors separate and unite rooms without limiting the living areas. In a part of the country known for its big skies, where logging is a way of life for many, houses constructed of glass and exposed lumber with simple tilted roof planes are a common sight. A compact building form is easier to keep comfortable in cold weather because exterior surfaces are kept to a minimum. Emory's design incorporates features of passive solar design, such as south-facing glass and structural insulated panel systems (SIPS) construction, for a house that is both energy efficient and low-maintenance.

Jeanie and Glenn purchased FabCab at the Seattle Home Show and had it re-assembled on a knoll where the view goes on forever. It was the home's accessibility—easy and invisible—that attracted the buyers from the start, with Jeanie's mother and sister in wheelchairs. But it is the little luxuries—an entry bench, recycled-glass countertops, and radiant heat—that make the home a delight to live in every day. They have renamed the house "Beau" in tribute to its beauty.

Long vistas make the tiny house feel much larger. The low kitchen counter, at two heights, tops a refrigerator and a drawer-type dishwasher. Frosted-glass pocket doors allow the house to function as two rooms or one large area.

A LOUISIANA PROTOTYPE

BeauSoleil (Cajun for "sunshine") is the name given to this 800-sq.-ft. house designed by Professor Geoff Gjertson's students at the University of Louisiana's Lafayette campus. Sparked by a national competition among collegiate teams to design, build, and operate solar-powered houses, Geoff's students added their own criteria. They would create a house relevant to the regional Cajun culture and address real needs in the state—affordable to median-income Louisiana families, and accessible to empty-nesters.

Adapting to Cajun culture

Two local housing types drove the layout. The "Shotgun" is long and narrow, a string of rooms along a linear pathway to maximize daylight. The "Dog Trot" adds a strong cross-axis, aligning doors and windows on opposite walls for natural ventilation. With these two images in mind, the students turned their attention to the Cajun culture. In a region where cooking is raised to an art form the kitchen can get pretty hot, so this kitchen can be isolated from the rest of the house. In a population where entertaining is a way of life, they designed a transitional porch—a natural breezeway, connected to a wraparound deck for large gatherings. Two pivoting walls of sliding glass easily transform the porch to an extra indoor room.

A design for empty-nesters

The students convened a focus group to understand accessibility issues for empty-nesters. For those with low vision, the design uses frosted

Originally built on the University campus (top), Beausoleil was disassembled and rebuilt in Washington DC as part of the 2009 US Department of Energy Solar Decathlon.

A compact accessible kitchen with a built-in snack-bar makes a great space for both entertaining and cooking at Beausoleil.

Floor Plan

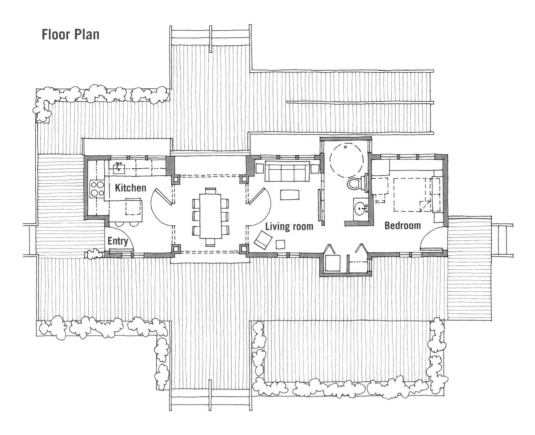

polycarbonate glass walls and doors to bring daylight through the house, and details that do not block the path of travel. For people with environmental sensitivities, the exterior building envelope (rainscreen siding systems and metal roofing) sheds water away from the house, while a dehumidification system limits the growth of mold. And for people with intellectual limitations, there are user-friendly mechanical and electrical systems, with remote controls and occupancy sensors for room lighting.

Adapting to Southern climate

The humid and hurricane-prone Southern climate added more design criteria. The plan cuts down on air-conditioning and heating by capturing natural breezes and sunshine. Roll-down screens offer insect control and shade protection. Rainwater collected from the roof fills a cistern for landscape irrigation. Walls built using SIPS stabilize indoor temperatures while resisting 130-mph winds. Operable steel shutters provide both window shade and hurricane protection. Long sunny days allow rooftop solar collectors and photovoltaic systems to generate more energy than the house consumes.

These two houses offer eloquent testimonial regarding the design potential for prefabricated housing to meet accessibility needs. Both have generous bathrooms with roll-in showers. Both have compact kitchens, where workspaces, storage, and appliances are all within easy reach. Both have shared living and dining rooms for flexibility in space use. With small square-footages, both avoid wasted space, to provide homes that are compact and easy to maintain and operate. This is the future of accessible prefab. ✦

The accessible bathroom features a low sink with knee space below, lever-handled plumbing controls, a hand-shower above a roll-in alcove, and a high toilet seat.

afterword: make it happen!

"Meeting the family and becoming part of their journey were the real rewards. I thought I was there to help them. In the end, they were the ones that gave me the lessons on how to make the most of every moment."

—WARREN RALSTON, ARCHITECT

Fortified with ideas and inspiration, making the leap to actually starting a construction project seems a huge undertaking. Many people spin their wheels for years before living conditions become unbearable—cabinet doors are falling off, light fixtures have gone dark, and the stair railing is tied with string and a prayer. A renovation project is a good time to add home-access upgrades because when disability occurs, decisions need to be made quickly, adding stress to an already-difficult adjustment period.

SELECT AN EXPERIENCED DESIGNER

Select a designer with the same care you'd use in selecting any other trusted professional, as you're relying on this person to keep your family safe and manage your investment. Seek recommendations from professional organizations and from friends who have had work done; it's just as useful to hear success stories as frustrations.

For finite projects such as a kitchen remodeling, a cabinetry supplier can be helpful. Interior and kitchen designers offer expanded services, including lighting, appliances, and finish selections. If you know just what you want, starting with

a builder may be appropriate. But when kitchen planning involves moving walls, adding a porch, and phasing future projects, architects are best equipped to orchestrate the work. Many states require drawings to be stamped and signed by a licensed architect, so ask your town's building inspector before hiring an unlicensed designer. An accessible home spans the boundaries between areas of specialization. For these reasons, the design process described in this book is architectural.

Draw up a shortlist of design firms to interview by checking websites and conducting phone interviews. Set up interviews with a few architects at your home so they can see how you live. Ask for references and also to see examples of their work. All things being equal, your choice should be based on your comfort levels. What is the designer's experience with creating accessible buildings? Does he listen to you? Do you feel confident she can represent your needs accurately? How will the designer work with you?

As fees are a reflection of services, any cost difference between two firms usually reflects different approaches. For an "apples-to-apples" comparison, take the time to understand what is being offered. Once you have chosen your designer, obtain a professional services agreement that itemizes responsibilities: what the designer will do, what may be done by others, what are basic vs. extra services, how you will be charged, what your role will be. Clearly written agreements reduce the chances of conflict.

Whether you can enlarge your home requires some investigation, starting with a trip to the local building department. Check zoning requirements, as expanding may require a variance. You may need a surveyor to see precisely how much space you have to expand on the site. A house near wetlands or in a historic district requires special approvals. If there are old drawings or files on record, having a copy made will save time and money later. These are tasks your architect can handle for you.

Early drawings are conceptual diagrams, general outlines of activity areas and spatial relationships. As the project progresses the drawings become more and more specific: Product selections and countless details are spelled-out in great detail. Final drawings (graphic representations) and specifications (product selection and quality standards) will be used for competitive pricing, for the building permit, and as the basis of the owner's contract with a builder. The more thorough these documents are, the more accurate will be the pricing and the fewer surprises there will be during construction.

PLAN TO BE A CONSTRUCTIVE MEMBER OF THE DESIGN TEAM

You want to be sure you are happy with the choices being made, so take as active a role as you are able (see the sidebar at right). Observe yourself in daily activities to understand your adaptations. Measure your comfort levels regarding workspace heights and storage reach ranges. This essential information forms the basis of the design.

- **Articulate your goals.** Identify problems; the architect's job is to identify solutions. What do you like about your home? What needs to change? Honesty about your needs and budget is a prerequisite to the designer being able to help you.
- **Be engaged.** Research on the Internet, try out products at other homes and public buildings, and visit showrooms.

Every product represents a deliberate choice. Knowing what you want will help you get it.

- **Communicate.** If the project is moving in a direction that concerns you, speak up. Ask questions. There's no way you should know the answers, especially if this is your first experience with construction.

SHOPPING STRATEGIES

A recurrent theme in speaking with homeowners is the difficulty of purchasing appliances when the homeowner has a disability. Online shopping is seldom detailed enough to visualize controls. On-site shopping has its own frustrations—long waits and underqualified staff. There is no real substitute for actually trying out the products you will be using regularly. You want handles that are comfortable to touch and controls that are easy to operate. Many people have settled for appliances they didn't want because the items could not be returned. Take the time to make informed decisions:

- **Do your research.** Use the Internet to identify products and features. Prioritize your requirements. Collect information and take notes.
- **Call ahead.** Find showrooms to visit. When you find a helpful salesperson, set a time to shop when they will be there.
- **Plug in.** Find a store that will let you plug in a floor model so you can see how clear the buttons or lights are, how the beeps sound, how easily the dials turn.
- **Take notes** on operation and maintenance, talking with sales staff and the manufacturer's representative. Review the manual together.
- **Find out what the return policy is.** If a store is unwilling to let you try the controls but charges a re-stocking fee, go to another store.

SELECT A CAPABLE GENERAL CONTRACTOR

Once construction starts, the general contractor (GC) coordinates the work of subcontractors and suppliers. Choosing a GC is like choosing an architect—and a roommate, as you'll essentially be living with him for a while. There are two basic ways to choose a GC. Competitive bidding lets you compare several proposals. A negotiated contract ensures you can work with the GC of your choice. Each method controls for different variables: cost, workmanship, approach, and schedule. Your architect can help you decide which method works best for you.

The contract itself has a bearing on the final price and should be part of your discussions with each GC. A lump-sum contract fixes the price up front, based on drawings and specs. Any changes made later will result in change-orders. This method is ideal for the owner who wants to know where the money is going, or to scale down the work before costs escalate past the budget. It does require thorough written description of the project, including drawings and specs.

Another option is the cost-plus-fee, where the GC charges directly for time and materials, plus either a lump sum, or a percentage of the cost of work, for overhead and profit. This method keeps the final cost a mystery until the work is done, but gives the owner the freedom to make decisions as the project unfolds. Adding a guaranteed maximum price (GMP) clause incentivizes the GC to work with the homeowner to keep costs from spiraling out of control. For both contractor selection and the owner-builder agreement, your architect can help navigate these decisions and tailor the process to your requirements.

AS A BUILDER, BE WILLING TO LEARN FROM YOUR CLIENTS

Accessible homes are a new idea in home-building, and old assumptions do not always adapt. Subs are likely to install electrical switches too high and suppliers to provide doors too narrow. "Measure twice, cut once," is the builder's mantra. When it comes to creating an accessible home, the stakes are higher, as people are less able to adapt. Measure three times—and check in often with the client.

- **Advise your subcontractors.** Accessible homes require special attention. Left-handed toilet flush valves won't work if the transfer space is on the right. Thresholds will need to be set on the sheathing rather than finish floors for a level walkway. These and other adaptations require a different focus for the construction team.
- **Build mock-ups.** One size does not fit all. Help the homeowner visualize and experiment with mounting heights for various items such as electrical fixtures or countertops. Help the user measure reach ranges to locate items correctly.
- **Control quality and workmanship.** Reconstruction is intrusive, messy, overwhelming, and costly. Get it right the first time.

KEEP THE DESIGN TEAM INVOLVED DURING CONSTRUCTION

During construction, the architect's involvement helps ensure that the work done reflects the design intent and the contract. The architect authorizes owner payments to the GC to prevent overspending. Routine job meetings allow the architect to track progress of the work and respond to hidden conditions. They troubleshoot problems and catch errors when corrections can be made. The architect provides design continuity to ensure the owner gets what was promised.

AS AN ARCHITECT, THINK OUTSIDE THE BOX

The architects who designed the homes in this book often describe their experiences as transformative. Designing an accessible home requires that the designer really understand what makes each client unique—how a family spends time together when one person has a disability, for instance, or how an illness affects other members of the household.

- **Ask questions.** Talk with the homeowner's occupational and physical therapists, caregivers, and family— people who understand how best to allow independence and provide a safe environment.

- **Be patient and resourceful.** Decisions may take a little longer when the homeowner has limited energy due to an illness or disability. Repurpose products from other settings. Modify institutional products for a residential setting.

- **Challenge yourself.** Bring design excellence to all aspects of the accessible home, and in so doing, raise the bar for all of us. The built environment, our clients with and without disabilities, and our profession deserve nothing less.

In looking with fresh eyes at the unique requirements of each person who will be in the home, architects are liberated from habitual ways of thinking about buildings. By looking closely at a person's disability, one sees clearly their abilities. It is in this spirit of seeing the whole person, and of transcending old ideas about housing, that the accessible home is created.

There is a misconception that architectural services are limited to producing drawings, but it is the architect's purposeful artistry and professionalism on behalf of the client's interests that adds value to a project. Design is both the product and the process, the harmonious blending of wishes and realities to create an environment that supports each person. When architects bring their best to this task, they serve homeowners and the profession.

THE ROLE OF THE ARCHITECT

Architects are designers, but they are also artists, technicians, and managers. Each project is a unique response to a unique set of circumstances. These include client needs, budgets and timetables, building code and local zoning requirements, building and site conditions, interior and exterior construction, engineering systems and services, built-in furnishings, plumbing and electrical fixtures, finishes and materials, and builder capabilities. In the process of creating the drawings, architects choreograph a team of specialists to ensure that the work of each is coordinated. These specialists may include engineers, lighting and acoustic consultants, interior and landscape designers, and others.

credits

Architect: Rob Whitten, Whitten Architects, Portland, ME; www.whittenarchitects.com
Contractor: Eider Construction, Scarborough, ME
Accessibility consultants: Alice and Jerry Holt

CASE STUDY 20
Home for Two Families
(pp. 195-198; and p. 46)
Designer: Bill Austin, Austin Design, Colrain, MA; www.austindesign.biz

CASE STUDY 21
Family Compound
(pp. 199-205; and pgs. 12 [left], 13 [left], 15, 22 [left, right], 28, 38, 41, 68 [top], 72, 77 [top], 92 [bottom], 94 [bottom])
Architect: B. H. Bocook AIA Architect, Palo Alto, CA; www.bocookarchitect.com
Contractor (Douglas's house only): Keith Alward, Alward Construction, Berkeley, CA; www.alwardconstruction.com

CASE STUDY 22
Reconstructed Barn
(pp. 207-212; and pgs. 34, 36, 53 [right], 58, 79 [bottom], 86)
Architect: Russell Campaigne, AIA, Mary Jo Kestner, AIA, CK Architects, Guilford, CT; www.ck-architects.com
Contractor: Mohovich Design & Carpentry, New Haven, CT

CASE STUDY 23
Aging in Place without Adding Space
(pp. 213-215; and pgs. 27 [right], 92 [top])
Architect: Ralph Rorem Architect, Kankakee, IL; www.ralphrorem.com
Contractor: Carl Clausen, Clausen Construction, Kankakee, IL

CASE STUDY 24
Accessible Tower
(pp. 216-221; and pgs. 29, 32, 33 [bottom], 40, 53 [left], 69 [top], 76, 79 [top], 84 [top])
Architect: Ross Chapin, FAIA, Ross Chapin Architects, Langley, WA; www.rosschapin.com
Contractors: George Piano, Phoenix Construction, Lake Forest Park, WA; www.myphoenixconstructioninc.com; Jim Soules, The Cottage Company, Seattle, WA

CASE STUDY 25
Prefab and Modular
(pp. 222-225)
FabCab architect: Emory Baldwin, AIA, Seattle, WA; www.fabcab.com
Contractor: Artisan Technologies, Roslyn, WA
BeauSoleil design team, contractor: University of Louisiana students (faculty advisor, W. Geoff Gjertson, AIA); www.beausoleilhome.org

OTHER HOUSES FEATURED
pp. 2-3, Architect: Michael Graves & Associates, Princeton, NJ; www.michaelgraves.com
p. 11 (bottom left), Architect: Reggie Stump, Astigmatic Studio, San Francisco, CA; www.astigmaticstudio.com
p. 19 (top); 93 (bottom), Architect: Russell Campaigne, AIA, CK Architects, Guilford, CT; www.ck-architects.com
p. 19 (bottom); 71 (left), Architect: House + House Architects, San Francisco, CA; houseandhouse.com
p. 20 (center right), Architect: Jessica Pappas, Fusco, Shaffer & Pappas, Farmington Hills, MI; www.fuscoshafferpappas.com
p. 21, p. 23 (top right), Architect: Catherine Roha, AIA, Berkeley, CA
p. 33 (top), Architect: LDa Architecture + Interiors, Cambridge, MA
p. 44, Designer: Lisa Bonneville, Bonneville Design, Manchester, MA
p. 50, Architect: Katy Flammia, AIA, THEREdesign, Boston, MA; theredesign.com

photo credits
All photos by Kathy Tarantola except as noted below:

p. 2-3: courtesy Clark Realty Capital
p. 8: Mark Hunt, Huntstock.com
p. 10: Kathryn Barnard, Photographer
p. 11: (left) Jerry Butts
p. 12: (right) Dale Lang
p. 19: (top right) Russell Campaigne; (bottom left) courtesy House + House Architects
p. 20: (center right): Christopher Lark, Inc
p. 23: (top left): Chris Green, courtesy *Fine Homebuilding* magazine, © The Taunton Press
p. 23: (top right) Dale Lang, nwphoto.net
p. 23: (bottom right) Deborah Pierce
p. 27: (left) Kathryn Barnard, Photographer
p. 33: (top) courtesy Greg Premru Photography Inc.
p. 35: (bottom) Dawn Connors Photography
p. 50: Katy Flammia
p. 57: (left) Kathryn Barnard, Photographer
p. 64: (top) Bob Handelman Images
p. 65: (bottom) Chris Green, courtesy *Fine Homebuilding* magazine, © The Taunton Press
p. 66: Martin Tessler
p. 69: (bottom) Dawn Connors Photography
p. 70: Kathryn Barnard, Photographer
p. 71: (right) courtesy House + House Architects

p. 73: Dale Lang
p. 84: (bottom) Dale Lang
p. 91: (top) Dawn Conners Photography
p. 91: (bottom) Emory Baldwin
p. 93: (bottom) Russell Campaigne
p. 94: (top) Mark Hunt, Huntstock.com
p. 103: Chris Green, courtesy *Fine Homebuilding* magazine, © The Taunton Press
p. 104: courtesy Universal Designers & Consultants, Inc.
p. 105: Chris Green, courtesy *Fine Homebuilding* magazine, © The Taunton Press
p. 106: (top left) Chris Green, courtesy *Fine Homebuilding* magazine, © The Taunton Press; (bottom right) courtesy Universal Designers & Consultants, Inc.
p. 107: Chris Green, courtesy *Fine Homebuilding* magazine, © The Taunton Press
p. 108: courtesy Universal Designers & Consultants, Inc.
p. 116: Kathryn Barnard, Photographer
p. 120: Kathryn Barnard, Photographer
pp. 121-124: Martin Tessler
pp. 140-145: Dawn Conners Photography
pp. 146-148, 150-151: Dale Lang
p. 149: (left) Dale Lang; (center) Emory Baldwin; (right) Mark Hunt, Huntstock.com
pp. 152-153: Randy O'Rourke
pp. 165-167: John Cate
p. 175: Deborah Pierce
pp. 185-187: Bob Handelman Images
p. 222: (bottom left) Emory Baldwin; (bottom right) © Dale Lang: nwphoto.net
p. 223: © Dale Lang: nwphoto.net
p. 224: (top) © 2009 Philip Gould
p. 224: (bottom) Catherine Guidry
p. 225: Catherine Guidry

technical credits
The author wishes to acknowledge the following who have contributed their expertise to the preparation of this book:

Access code applications and jurisdictions: David Kessler, Kessler McGuinness & Associates, West Newton, MA; www.kmaccess.com
Architectural acoustics (p. 39): Greg Tocci, Cavanaugh Tocci Associates, Sudbury, MA; www.cavtocci.com
Lighting the home (p. 41): Nancy Goldstein, Nancy Goldstein Design, Marblehead, MA; www.ngdesign.net
Heating, Ventilating, and Air Conditioning (HVAC) systems (p. 78): Sergio Siani, Norian/Siani Engineering, Inc., Waltham, MA; www.ns-engineering.com

index